ADSORPTION
Surface Chemistry

P. Tushar

RAJAT PUBLICATIONS
New Delhi

Adsorption: *Surface Chemistry*
by P. Tushar

This edition is published by Rajat Publications
4740/23, Ansari Road Daryaganj,
New Delhi - 110002

ISBN: 978-81-7880-128-5

Edition 2026

Printed and bound in India

RAJAT
PUBLICATIONS

4740/23, Ansari Road, Daryaganj,
New Delhi- 110002 (India)
Tel:- +91-11-23267924
Mob:- +91-9811484973, +91-9811105389
E-mail: rajat_publication@yahoo.com

PREFACE

This text is a concise yet comprehensive text at the undergraduate and postgraduate levels, enabling easy access to the core information about adsorption (surface chemistry). This text covers all the important areas of adsorption (surface chemistry) in a format that is ideal for learning and rapid revision. Wherever necessary, line diagrams have been added to make the subject more clear. Throughout the text, the SI units have been employed. This text has been designed to cover the latest UGC syllabus. This text includes essential concepts to keep pace with new developments in adsorption (Surface Chemistry).

—Editor

Contents

1
Introduction

Adsorption

A boundary that separates two phases is known as a *surface* or an *interface*. Surfaces show special properties that are different from those of the phases themselves. For example, the surface of a solid often shows a strong affinity for molecules that come into contact with it and which are said to be ***adsorbed***. A simple way of visualising adsorption is in terms of additional valence bonds at the surface, which are available for bonding; in reality, however, the situation is more complex.

Adsorption is a process in which atoms or molecules move from a bulk phase (that is, solid, liquid or gas) onto a solid or liquid surface. An example is purification by adsorption where impurities are filtered from liquids or gases by their adsorption onto the surface of a high-surface-area solid such as activated charcoal. Other examples include the segregation of surfactant molecules to the surface of a liquid, the bonding of reactant molecules to the solid surface of a heterogeneous catalyst, and the migration of ions to the surface of a charged electrode.

Adsorption is to be distinguished from absorption, a process in which atoms or molecules move into the bulk of a porous material, such as the absorption of water by a sponge. Sorption is a more general term that includes both adsorption and absorption. Desorption refers to the reverse of adsorption, and is a process in which molecules adsorbed on a surface are transferred back into a bulk phase. The term adsorption is most often used in the context of solid surfaces in contact with liquids and gases. Molecules that have been adsorbed onto solid surfaces are referred to generically as adsorbates, and the surface to which they are adsorbed as the substrate or adsorbent. See also: *Absorption.*

Process

At the molecular level, adsorption is due to attractive interactions between a surface and the species being adsorbed. The magnitude of these interactions covers approximately two orders of magnitude (8-800 kilojoules/mole), similar to the range of interactions found between atoms and molecules in bulk phases. Traditionally, adsorption is classified according to the magnitude of the adsorption forces. Weak interactions (<40 kJ/mol) analogous to those between molecules in liquids give rise to what is called physical adsorption or physisorption. Strong interactions (>40 kJ/mol) similar to those found between atoms within a molecule (for example, covalent bonds) give rise to chemical adsorption or chemisorption. In physisorption the adsorbed molecule remains intact, but in chemisorption the molecule can be broken into fragments on the surface, in which case the process is called *dissociative chemisorption.*

The extent of adsorption depends on physical parameters such as temperature, pressure, and concentration in the bulk phase, and the surface area of the adsorbent, as well as on chemical parameters such as the elemental nature of the adsorbate and the adsorbent. Low temperatures, high pressures, high surface areas, and highly reactive adsorbates or adsorbents generally favour adsorption.

A thermodynamic description of adsorption quantifies the driving force for adsorption in terms of these parameters. An important quantity that can be defined in this regard is the Gibbs free energy of the surface, that is, the surface free energy. For a one- component system, surface free energy is also called *surface tension,* and it corresponds to the so-called reversible work required to create a surface. Surface free energy is the two-dimensional analog of pressure, and whereas pressure is measured as force per unit area, surface free energy is given as force per unit length. Common units for surface free energy are dynes/cm or, equivalently, ergs/cm^2. Easily created surfaces such as those for liquids and layered solids (for example, graphite and mica) have low surface free energies of 0-500 ergs/cm^2. Clean surfaces of tough, refractory materials have high free energies of up to several thousand ergs per square centimeter. At constant temperature and pressure, spontaneous processes lower the free energy of a system, and adsorption generally lowers the free energy of a surface; therefore, adsorption tends to occur most readily on materials with high surface free energies. See also: *Gibbs function.*

A common way to portray the results of adsorption studies on solid surfaces is in the form of an adsorption isotherm. Such a diagram gives the amount of adsorbed material per surface area at a constant temperature as a function of pressure or concentration in the bulk phase. Adsorption initially increases with pressure and then saturates at a value that corresponds to one layer (a monolayer) of adsorbed molecules. The adsorbed monolayer has a lower free energy than the clean surface, so the driving force for adsorption is decreased upon completion of the monolayer. At still higher pressures, however, the free energy of the molecules in the gas is high enough that there is measurable adsorption of molecules on top of the monolayer and multiple layers or multilayers can be condensed onto the surface.

An adsorption isotherm represents a chemical equilibrium in which molecules are simultaneously adsorbing to and desorbing from the surface. The isotherm shows the extent of the adsorbed layer that is the net result of these two competing processes. For many adsorption systems, particularly those involving physical adsorption, the potential energy decreases monotonically to a minimum as the adsorbate approaches the surface so that there is no energy barrier for adsorption, and equilibrium is readily achieved. Other systems involve activated adsorption, and the adsorbate encounters a potential energy barrier as it approaches the surface. This phenomenon is commonly encountered in dissociative chemisorption; the process can be depicted schematically in a one-dimensional potential diagram (Fig. 1.1). The height of the potential energy barrier in such a system is determined by the extent to which the molecular fragments bond to the surface as the bond within the molecule is being broken. See also: *Chemical thermodynamics.*

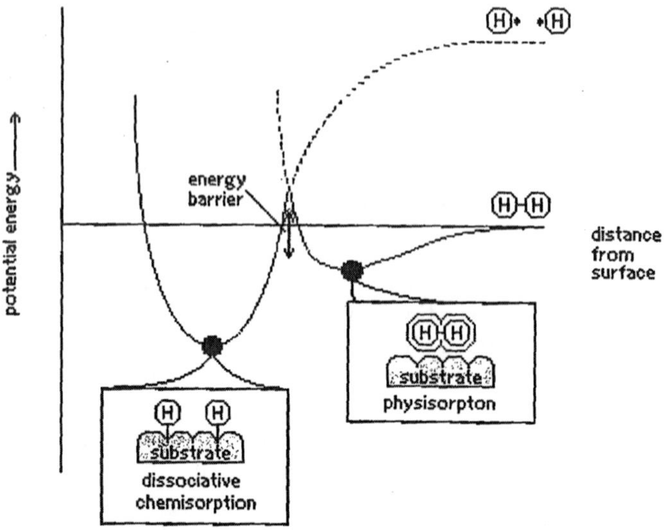

Fig. 1.1: One-dimensional potential energy curves typical for hydrogen molecules and atoms as they approach the surface of a metal substrate. The solid curve, which involves a crossing from the molecular curve to the atomic curve, shows the type of potential profile expected for dissociative chemisorption of hydrogen molecules on the surface.

Detection and Study

Studying adsorption, particularly in the monolayer regime, is difficult because of the small numbers of molecules involved. For 1 cm^2 of surface area, the number of surface atoms or adsorbed molecules in a monolayer is of the order of 10^{15}, which is only ~ 10^{-9} mole. Often, however, adsorbents have surface areas of hundreds of square centimeters per gram, and adsorption can be quantified simply by measuring the change in pressure or concentration of the bulk phase. Adsorption can also be quantified by measuring the change in mass of the adsorbent and by monitoring changes in the properties of the surface of the adsorbent such as the surface tension and the surface potential. By determining the number of molecules in an adsorbed monolayer, constructing an isotherm, and knowing the size of the adsorbed molecule, the area of the surface can be determined. Such a procedure is the basis for the Brunauer-Emmett-Teller (BET) method for determining surface areas.

Spectroscopies are commonly employed to study directly the structure and bonding of atoms or molecules in adsorbed monolayer. Spectroscopies that measure the absorption of light have the advantage of high resolution but the disadvantage of a low absorption cross section. High-intensity light sources (lasers and synchrotrons) and high-sensitivity detectors are often required to compensate for low absorption cross sections. If, however, the adsorbent does not interact strongly with the light, optical spectra of adsorbates can be obtained by using high-surface-area adsorbents.

Higher-sensitivity detection of adsorbed monolayers can be achieved by using electrons, ions, and atoms as probes. An advantage of these techniques is that because the probing particles do not significantly penetrate the solid, the information that they provide is surface-specific. Interactions of the probing particles with gas phase and liquid molecules, however, limit these types of studies to vacuum conditions where the particles can be scattered from adsorbed monolayers without interference from the gas phase. See also: *Monomolecular film.*

The application of spectroscopies to study adsorbed monolayers has formed the basis for what is known as *surface science*. A wide variety of spectrosopic and differaction techniques that utilise various combinations of electrons, photons, atoms, and ions as probes have been developed as part of surface science. These include Auger electron spectroscopy (AES), x-ray photoelectron spectroscopy (XPS), electron energy loss spectroscopy (EELS), ion scattering spectroscopy (ISS), and electron or atom diffraction.

Another vacuum-based technique that is frequently applied to probe the strength of surface-adsorbate bonds and to study chemical reactions in adsorbed monolayers is temperature-programmed desorption (TPD). In a temperature-programmed desorption experiment, a substrate covered with an adsorbed monolayer is heated, and the species that desorb from the surface are detected (typically with a mass spectrometer) as a function of surface temperature. The resulting temperature-programmed desorption spectrum consists of peaks that are proportional to the amount of material desorbed, and the peak temperatures are related to the strengths of the adsorbate-surface bonds—higher temperatures corresponding to stronger bonds. If a reaction occurs in the adsorbed monolayer during heating, a temperature-programmed desorption spectrum shows the products that are evolved from the surface and indicates the rate at which they are produced.

Many surface science studies are carried out under ultrahigh-vacuum conditions (pressure = 10^{-8} to 10^{-11} torr, or 10^{-3} to 10^{-6} pascal). Ultrahigh-vacuum conditions make it possible to prepare and maintain atomically clean surfaces of materials that react with air. Often in these studies, samples of single crystals are utilised, and surfaces that are atomically smooth over the range of thousands of atoms can be prepared. These well-defined surfaces can serve as model systems for understanding the molecular details of adsorption.

A general finding from surface science studies of adsorption on single crystal surfaces is that, even for strongly chemisorbed layers, the adsorbate bonding to the surface is directly analogous to ligand bonding in discrete molecular compounds. An example is the ethylidyne ligand (CCH_3) bound to a rhodium surface and to a molecular cobalt complex (Fig. 1.2)

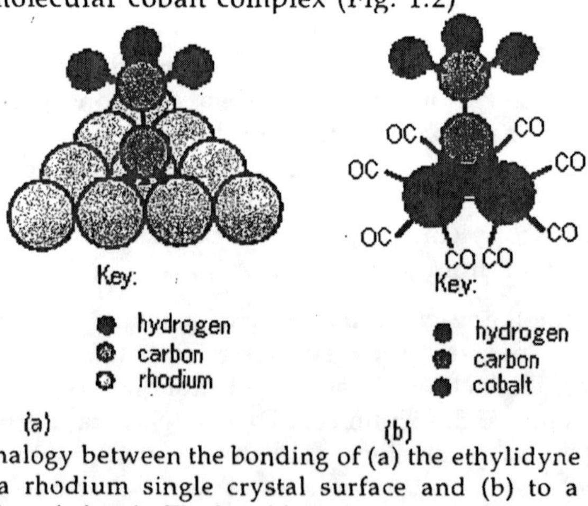

Fig. 1.2: Analogy between the bonding of (a) the ethylidyne ligand to a rhodium single crystal surface and (b) to a cobalt carbonyl cluster. The bond lengths and angles are virtually identical in the two systems.

An exciting frontier in surface science studies of adsorption is the application of scanning tunneling microscopy (STM). This technique can provide an atomic resolution image of the electronic structure at the surface. See also: *Scanning tunneling microscope.*

The adsorption of molecules on to a surface is a necessary prerequisite to any surface mediated chemical process.

For example, in the case of a surface catalysed reaction it is possible to break down the whole continuously-cycling process into the following five basic steps:

1. Diffusion of reactants to the active surface
2. Adsorption of one or more reactants onto the surface
3. Surface reaction
4. Desorption of products from the surface
5. Diffusion of products away from the surface

The above scheme not only emphasises the importance of the adsorption process but also its reverse namely desorption. It is these two processes which are considered in this Section.

Notes on Terminology

Substrate - Frequently used to describe the solid surface onto which adsorption can occur; the substrate also occasionally (although not here) referred to as the adsorbent.

Adsorbate - The general term for the atomic or molecular species which are adsorbed (or are capable of being adsorbed) onto the substrate.

Adsorption - The process in which a molecule becomes adsorbed onto a surface of another phase (note - to be distinguished from absorption which is used when describing uptake into the bulk of a solid or liquid phase).

Coverage - A measure of the extent of adsorption of a species onto a surface (unfortunately this is defined in more than one way !). Usually denoted by the lower case Greek "theta", θ.

Exposure - a measure of the amount of gas which as surface has seen; more specifically, it is the product of the pressure and time of exposure (normal unit is the Langmuir, where 1 L = 10^{-6} Torr s).

Applications

Direct applications of adsorption in processes such as filtration and detergent action are well known. Adsorption is also the basis for a series of vacuum pumps known as *getter pumps*. In these pumps, molecules are removed from the gas phase either by physisorption on high-surface-area materials at low temperatures (cryopumps and sorption pumps) or by chemisorption on highly reactive metal surfaces (ion pumps and titanium sublimation pumps).

Adsorption also plays an important role in processes such as heterogeneous catalysis, electrochemistry, adhesion, lubrication, and molecular recognition. In heterogeneous catalysis, gas or solution-phase molecules adsorb onto the catalyst surface, and reactions in the adsorbed monolayer lead to products which are desorbed from the surface. In electrochemistry, molecules adsorbed to the surface of an electrode donate or accept electrons from the electrode as part of oxidation or reduction reactions. In adhesion and lubrication, the chemical and mechanical properties of adsorbed monolayers play a role in determining how solid surfaces behave when in contact with one another. In biological systems, the adsorption of atoms and molecules onto to the surface of a cell membrane is the first step in molecular recognition. See also: *Electrochemistry; Heterogeneous catalysis; Molecular recognition.*

Because of the many important technological applications of adsorption, studies of adsorption are important in disciplines ranging from solid-state physics and physical chemistry to materials science and molecular biology.

Sorption

Sorption, a term used to denote the separation of a chemical species between bulk and interface, is frequently encountered in surface chemistry. We also differentiate between changes that occur on the external surface and those which involve the interior. Species are adsorbed on the outside and absorbed on the inside. Take the case of a porous solid—it is rather like a sponge, but the inside structure is very much smaller; many examples occur in nature, and they are also manufactured for specific applications. The pores have dimensions of the order of 1—100 nm, and gases are absorbed into them; the concentration of gas in the pore is greater than that in the bulk phase and it is said to be *positively* absorbed. Adsorption occurs on the external surface of the solid, and again the gas is *positively adsorbed.* Negative adsorption is also possible although less common. It occurs in electrolyte solutions when the concentration of ions at the surface is less than that in the bulk solution.

Surface Groups

Because the solid has a particular bulk chemical composition, it does not necessarily follow that the surface contains the same atoms arranged in a similar way to the bulk phase. Often as the result of chemical interactions with the components of the surrounding gas phase, a number of different chemical groups may be found on the surface. The best understood examples of this are metallic oxides, silica and carbons.

As an example of the behaviour of metal oxides, rutile presents a typical case. The changes in the rutile spectra caused by heating and evacuation were explained by Jackson and Parfitt in terms of the bridged and terminal hydroxyl groups formed by the dissociation of adsorbed water on the (110) plane.

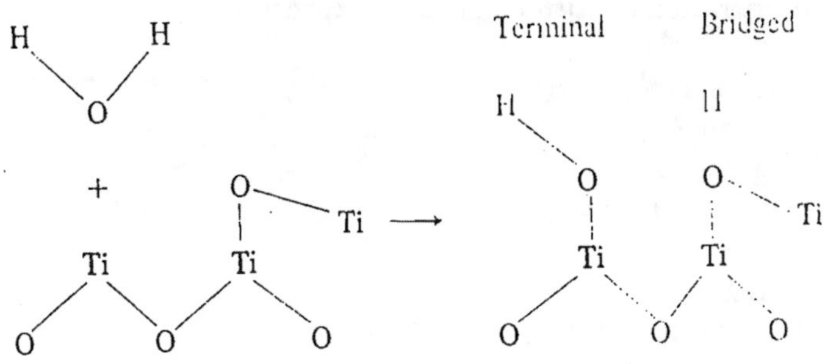

Hockey and Jones found differences in the behaviour of (100), (101) and (110) planes, suggesting that the first two absorb molecular water as a ligand co-ordinated to a surface Ti^{4+} ion, whereas the adsorption of water on the (110) plane involves the dissociation of the water molecule and the formation of two differently bonded OH^- ions. Calculations by Waldsax and Jaycock suggest that the water molecule would adsorb preferentially on the five co-ordinate Ti^{4+} sites in this plane, and then dissociate to give the two types of hydroxyl group.

A silica particle can be regarded as a polymer of sillicic acid, which terminates at the surface in either a siloxane grouping, ⋟Si—OSi⋞, with the oxygen in the surface, or as the result of interaction with water, one of several forms of silanol group, ⋟Si—OH. These can be defined as follows:

> isolated — where the surface Si atom has one bond attached to the —OH group which is not hydrogen bonded to another silanol group;
>
> vicinal — where two single silanol group attached to different Si atoms are close enough to hydrogen bond to each other; and
>
> geminal — where two —OH groups attached to the same Si atom, leaving only two bonds from the Si atom to the bulk structure, which are too close to hydrogen bond to each other.

Commercial carbon blacks are usually found to possess five major chemical groups on their surface; aromatic hydrogen, phenol, quinone, carboxylic acid and lactone. Rivin found for a furnace black with a surface area of 76 m^2g^{-1} the following concentration of surface groups, expressed in milli-equivalents g^{-1}:

aromatic hydrogen	1.94	(0.065)
phenol	0.61	(0.21)
quinone	0.00	
carboxylic acid	0.01	(12)
lactone	0.17	(0.74)

where the figures in brackets represent the area of surface per group in nm^2. It is possible that some of the hydrogen may not in fact be in the surface but may penetrate the bulk of the carbon.

2

Types of Adsorption

Introduction

It is now recognised that there are two main types of adsorption. In the first type, the forces are of a physical nature and the adsorption is relatively weak. The forces correspond to those considered by J.H. van der Waals in connection with his equation of state for gases and are known as *van der Waals* forces. This type of adsorption is known as *physical adsorption, physisorption,* or *van der Waals adsorption*. The heat evolved when a mole of gas becomes physisorbed is usually small, less than 20 kJ. This type of adsorption plays only an unimportant role in catalysis, except for certain special types of reactions involving free atoms or radicals.

In the second type of adsorption, first considered in 1916 by the American chemist Irving Langmuir (1881-1957), the adsorbed molecules are held to the surface by covalent forces of the same general type as those occurring between atoms in molecules. The heat evolved per mole for this type of adsorption, known as *chemisorption*, is usually comparable to that evolved in chemical bonding, namely 100 to 500 kJ mol^{-1}.

An important consequence of chemisorption is that after a surface has become covered with a single layer of adsorbed molecules, it is *saturated;* additional adsorption can occur only on the layer already present, and this is generally weak adsorption. Langmuir thus emphasised that chemisorption involves the formation of a *unimolecular layer.* Many investigations on surfaces of known area have confirmed that chemisorption ceases after a unimolecular layer is formed, but that physisorption may give rise to additional layers.

Langmuir's 1932 Nobel Prize for chemistry was for his discoveries in the realm of surface chemistry.

It was suggested in 1931 by Hugh Stott Taylor (1890-1974) that chemisorption is frequently associated with an appreciable activation energy and may therefore be a relatively slow process. For this reason chemisorption is often referred to as *activated adsorption*. By contrast, van der Waals adsorption

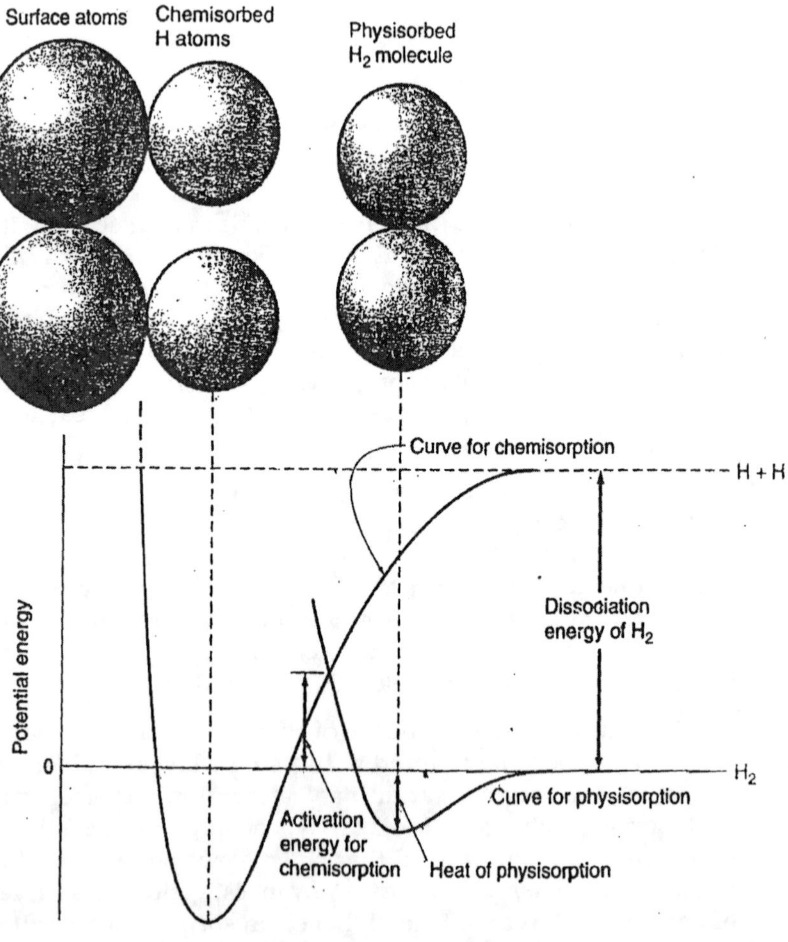

Fig. 2.1: Schematic potential-energy curves for the physisorption and chemisorption of hydrogen on a surface.

requires no activation energy and therefore occurs more rapidly than chemisorption.

The relationship between physisorption and chemisorption is illustrated by the potential-energy diagram shown in Figure 2.1. It relates to the adsorption of hydrogen on a surface such as that of a metal. Physisorption occurs first, the intact molecule being held loosely to the surface by dispersion forces. The potential-energy curve for chemisorption, with dissociation of the hydrogen molecule, has a much deeper minimum, corresponding to stronger bonding. The crossing of the two curves shows that there is a small energy of activation for the chemisorption process. This diagram shows that initial physisorption is an important feature of the chemisorption process. If there were no physisorption, there would be a much higher activation energy for the chemisorption; the process via the physisorbed stage is therefore always favoured.

Another important concept, suggested in 1925 by H.S. Taylor, is that solid surfaces are never completely smooth and that adsorbed molecules will be attached more strongly to some surface sites than to others. This is particularly important in connection with catalysis, since chemical reaction may occur predominantly on certain sites, which Taylor referred to as *active centres*.

How do Molecules Bond to Surfaces?

There are two principal modes of adsorption of molecules on surfaces:

Physical Adsorption (Physisorption)

Chemical Adsorption (Chemisorption)

The basis of distinction is the nature of the bonding between the molecule and the surface. With...

Physical Adsorption: The only bonding is by weak Van der Waals - type forces. There is no significant redistribution of electron density in either the molecule or at the substrate surface.

Chemisorption: A chemical bond, involving substantial rearrangement of electron density, is formed between the adsorbate and substrate. The nature of this bond may lie anywhere between the extremes of virtually complete ionic or complete covalent character.

Typical Characteristics of Adsorption Processes

	Chemisorption	*Physisorption*
Temperature Range (over which adsorption occurs)	Virtually unlimited (but a given molecule may efectively adsorb only over a small range)	Near or below the condensation point of the gas (e.g. Xe < 100 K, CO_2 < 200 K)
Adsorption Enthalpy	Wide range (related to the Chemical bond strength)- typically 40 - 800 kJ mol^{-1}	Related to factors like molecular mass and polarity but typically 5-40 kJ mol^{-1} (i.e. ~ heat of liquefaction)
Crystallographic Specificity(variation between different surface planes of the same crystal)	Marked variation between crystal planes	Virtually independent of surface atomic geometry
Nature of adsorption	Often dissociative May be irreversible	Non—dissociative Reversible
Saturation Uptake	Limited to one monolayer	Multilayer uptake possible
Kinetics of Adsorption	Very variable - often an activated process	Fat-since it is a non-activated process

The most definitive method for establishing the formation of a chemical bond between the adsorbing molecule and the substrate (i.e. chemisorption) is to use an appropriate spectroscopic technique, for example,

IR - to observe the vibrational frequency of the substrate/ adsorbate bond.

UPS - to monitor intensity & energy shifts in the valence orbitals of the adsorbate and substrate.

3

Adsorption Kinetics - The Rate of Adsorption

The rate of adsorption of a molecule onto a surface can be expressed in the same manner as any kinetic expression: i.e.

$R_{ads} \ k\ C^{x}$ where: x - kinetic order

k - rate constant

C - gas phase concentration

or

$R_{ads} = k'\ P^{x}$ where: x - kinetic order

k' - rate constant

P - partial pressure of molecule

If the rate constant in either of the above expressions is expressed in an Arrhenius form, then we obtain kinetic equations of the form:

$$R_{ads} = A\ C^{x}\ \exp(-E_a / RT)$$

where E_a is the *activation energy for adsorption*, and A the *pre-exponential (frequency) factor*.

It is much more informative, however, to consider the factors controlling this process at the molecular level . . . .

The rate of adsorption is governed by

1. the rate of arrival of molecules at the surface
2. the proportion of incident molecules which undergo adsorption

i.e. we can express the rate of adsorption (per unit area of surface) as a product of the incident molecular flux, F, and the sticking probability, S.

$R_{ads} = S.\ F$ [molecules $m^{-2}\ s^{-1}$]

The flux of incident molecules is given by the Hertz-Knudsen equation

Flux, $F = P\ /\ (2\pi mkT)^{1/2}$ [molecules $m^{-2}s^{-1}$]

where

p - gas pressure [N m^{-2}]

m - mass of one molecule [kg]

T - temperature [K]

The sticking probability is clearly a property of the adsorbate/substrate system under consideration but must lie in the range $0 < S < 1$; it may depend upon various factors - foremost amongst these being the existing concentration of adsorbed species (n) and the presence of any activation barrier to adsorption. In general, therefore,

$S = f\ (\theta).\ \mathrm{Exp}\ (-E_a/RT)$

where, once again, E_a is the activation energy for adsorption and f (θ) is some, as yet undetermined, function of the existing surface coverage of adsorbed species.

Combining the equations for S and F yields the following expression for the rate of adsorption:

$$R = \frac{f\ (\theta)\ .\ P}{\sqrt{2\pi mkT}}\ \exp\ (-E_a\ /\ RT)$$

Note:

1. It should be recognised that the activation energy for adsorption may itself be dependent upon the surface coverage, i.e. $E_a = E(\theta)$.

2. If it is further assumed that the sticking probability is directly proportional to the concentration of vacant surface sites (which would be a reasonable first

assumption for non-dissociative adsorption) then $f(\theta) = (1-\theta)$; where, in this instance, θ is the fraction of sites which are occupied (i.e. the Langmuir definition of surface coverage).

Estimating Surface Coverages arising as a result of Gas Exposure

If a surface is initially clean and it is then exposed to a gas pressure under conditions where the rate of desorption is very slow, then the coverage of adsorbed molecules may initially be estimated simply by consideration of the kinetics of adsorption.

As noted above, the rate of adsorption is given by : $R_{ads} = S.F$

i.e.

$$\frac{dN_{ads}}{dt} = SF$$

where : N_{ads} is the number of adsorbed species per unit area of surface.

In general, this equation must be integrated to obtain an expression for N_{ads}, since the sticking probability is coverage (and hence also time) dependent.

However, if it is assumed that the sticking probability is essentially constant (which may be a reasonable approximation for relatively low coverages), then this integration simply yields;

$$N_{ads} = SFt$$

4

PE Curves & Energetics of Adsorption

In this section, we will consider both the energetics of adsorption and factors which influence the kinetics of adsorption by looking at the "potential energy diagram/curve" for the adsorption process.

The potential energy curve for the adsorption process is a representation of the variation of the energy (PE or E) of the system as a function of the distance (d) of an adsorbate from a surface.

Within this simple one-dimensional model, the only variable is the distance (d) of the adsorbing molecule from the substrate surface. Thus, the energy of the system is a function only of this variable i.e. $E = E(d)$

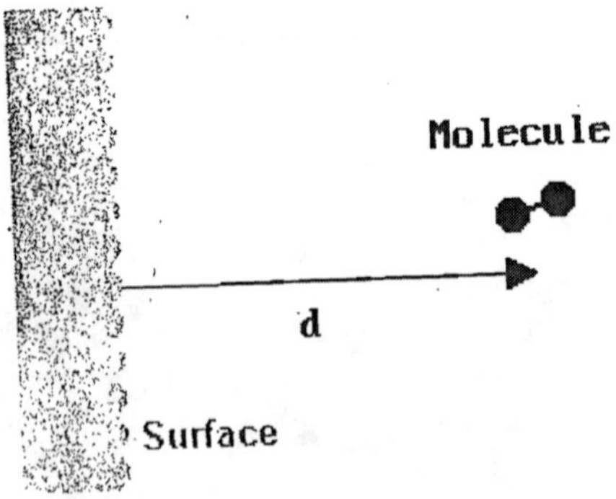

Fig. 4.1

It should be remembered that this is a very simplistic model which neglects many other parameters which influence the energy of the system (a single molecule approaching a clean surface), including for example

- *the angular orientation of the molecule*
- *changes in the internal bond angles and bond lengths of the molecule*
- *the position of the molecule parallel to the surface plane*

The interaction of a molecule with a given surface will also clearly be dependent upon the presence of any existing adsorbed species, whether these be surface impurities or simply pre-adsorbed molecules of the same type (in the latter case we are starting to consider the effect of surface coverage on the adsorption characteristics).

Nevertheless it is useful to first consider the interaction of an isolated molecule with a clean surface using the simple ID model. For the purposes of this tutorial, we will also not be overly concerned whether the "energy" being referred to should strictly be the internal energy, the enthalpy or free energy of the system.

CASE I - Physisorption

In the case of pure physisorption (e.g. Ar/metals), the only attraction between the adsorbing species and the surface arises from weak, van der Waals forces.

As illustrated below, these forces give rise to a shallow minimum in the PE curve at a relatively large distance from the surface (typically d > 0.3 nm) before the strong repulsive forces arising from electron density overlap cause a rapid increase in the total energy.

There is no barrier to prevent the atom or molecule which is approaching the surface from entering this physisorption well, i.e. the process is not activated and the kinetics of physisorption are invariably fast.

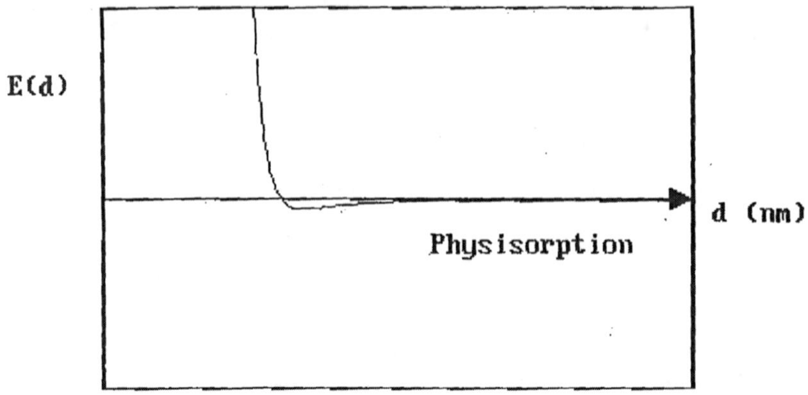

Fig. 4.2

CASE II - Physisorption + Molecular Chemisorption

The weak physical adsorption forces and associated long-range attraction will be present to varying degrees in all adsorbate / substrate systems. However, in cases where chemical bond formation between the adsorbate and substrate can also occur, the PE curve is dominated by a much deeper chemisorption minimum at shorter values of d.

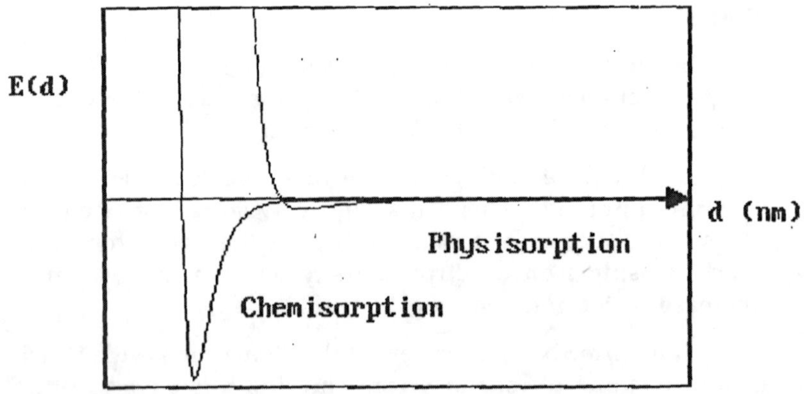

Fig. 4.3

The Figure 4.3 shows the PE curves due to physisorption and chemisorption separately—in practice, the PE curve for any real molecule capable of undergoing chemisorption is best described by a combination of the two curves, with a curve crossing at the point at which chemisorption forces begin to dominate over those arising from physisorption alone.

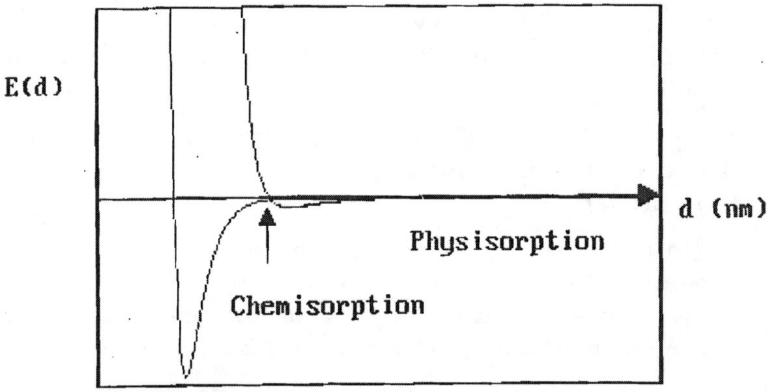

Fig. 4.4

The minimum energy pathway obtained by combining the two PE curves is now highlighted in red. Any perturbation of the combined PE curve from the original, separate curves is most likely to be evident close to the highlighted crossing point.

For clarity, we will now consider only the overall PE curve:

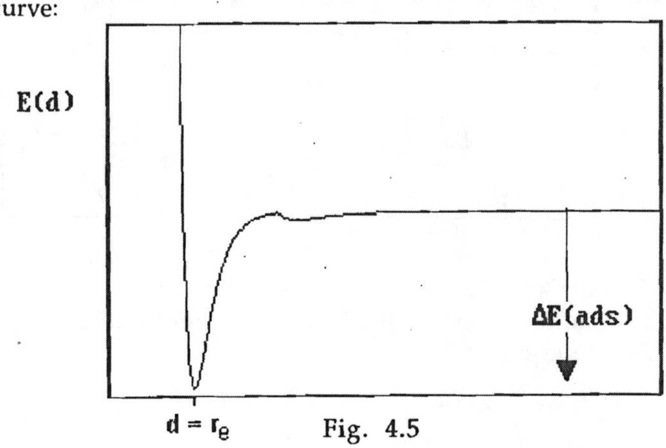

Fig. 4.5

The depth of the chemisorption well is a measure of the strength of binding to the surface - in fact it is a direct representation of the energy of adsorption, whilst the location of the global minimum on the horizontal axis corresponds to the equilibrium bond distance (r_e) for the adsorbed molecule on this surface.

The energy of adsorption is *negative*, and since it corresponds to the energy *change* upon adsorption it is better represented as ΔE (ads) or ΔE_{ads}. However, you will also often find the depth of this well associated with the enthalpy of adsorption, ΔH (ads).

(Note - the "*heat of adsorption*", Q, is taken to be a positive quantity equal in magnitude to the enthalpy of adsorption; i.e. Q = -ΔH(ads))

In this particular case, there is clearly no barrier to be overcome in the adsorption process and there is no activation energy of adsorption (i.e. $E_a^{ads} = 0$, but do remember the previously mentioned limitations of this simple ID model).

There is of course a significant barrier to the reverse, desorption process - the red arrow in the diagram below represents the activation energy for desorption.

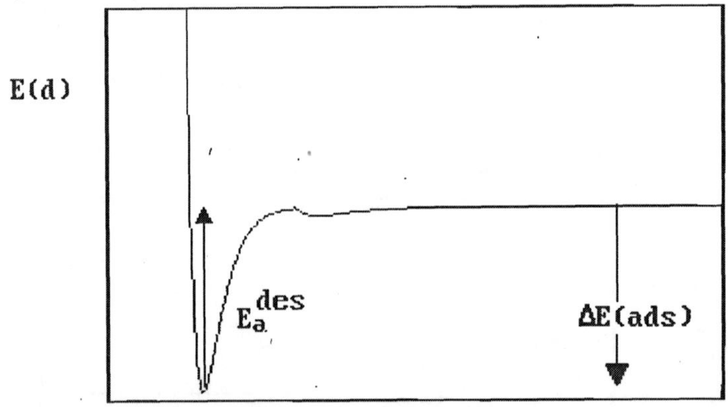

Fig. 4.6

Clearly in this particular case, the magnitudes of the energy of adsorption and the activation energy for desorption can also be equated i.e.

$E_a^{des} = \Delta E$ (ads) or $E_a^{des} \sim \Delta H$ (ads)

CASE III - Physisorption + Dissociative Chemisorption

In this case, the main differences arise from the substantial changes in the PE curve for the chemisorption process.

Again, we start off with the basic PE curve for the physisorption process which represents how the molecule can weakly interact with the surface:

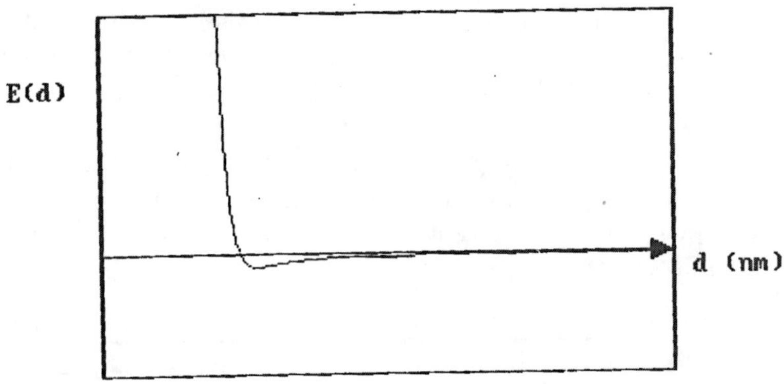

Fig. 4.7

If we now consider a specific molecule such as H_2 and initially treat it as being completely isoslated from the surface (i.e. when the distance, d, is very large) then a substantial amount of energy has to be put into the system in order to cause dissociation of the molecule.

$$H_2 \rightarrow H + H$$

- This is the bond dissociation energy [D (H-H)], some 435 kJ mol^{-1} or 4.5 eV.

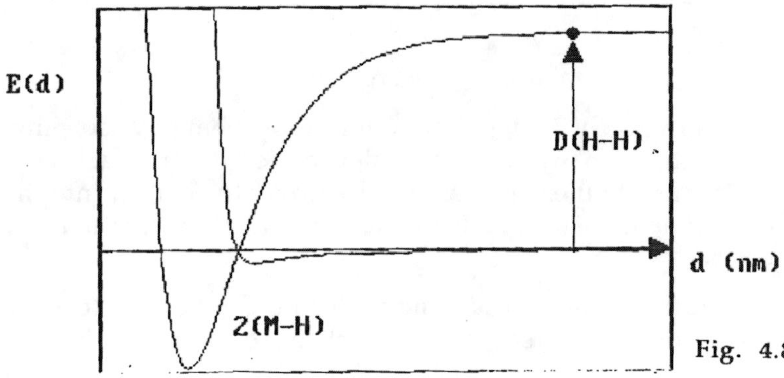

Fig. 4.8

The red dot in the Figure 4.8 thus represents two hydrogen atoms, equidistant (and a long distance) from the surface and also now well separated from each other. If these atoms are then allowed to approach the surface they may ultimately both form strong chemical bonds to the substrate...this corresponds to the minimum in the red curve which represents the chemisorption PE curve for the two H atoms.

In reality, of course, such a mechanism for dissociative hydrogen chemisorption is not practical - the energy down payment associated with breaking the H-H bond is far too severe.

Instead, a hydrogen molecule will initially approach the surface along the physisorption curve. If it has sufficient energy it may pass straight into the chemisorption well ("direct chemisorption")....

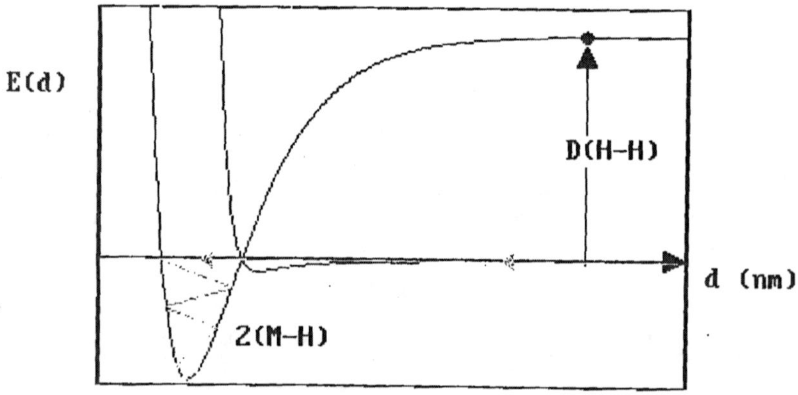

Fig.4.9

or, alternatively, it may first undergo transient physisorption —a state from which it can then either desorb back as a molecule into the gas phase or cross over the barrier into the dissociated, chemisorptive state (as illustrated *schematically* below).

Into this latter case, the molecule can be said to have undergone "precursor-mediated' chemisorption.

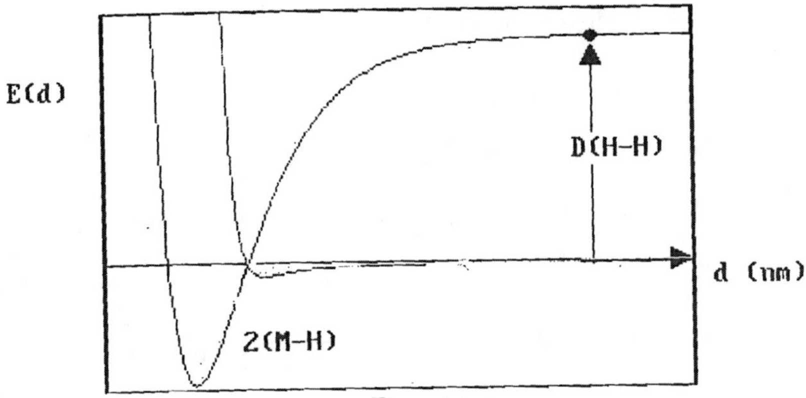

Fig. 4.10

The characteristics of this type of dissociative adsorption process are clearly going to be strongly influenced by the position of the crossing point of the two curves (molecular physisorption v's dissociative chemisorption) - relatively small shifts in the position of either of the two curves can significantly alter the size of any barrier to chemisorption.

In the example immediately below there is no direct activation barrier to dissociative adsorption - the curve crossing is below the initial "zero energy" of the system.

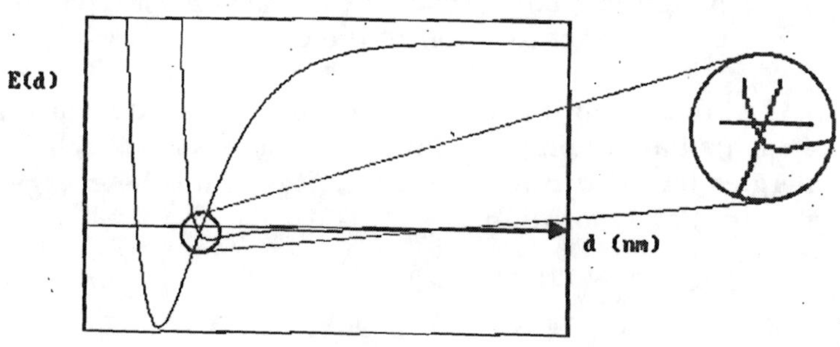

Fig. 4.11

whilst, in this next case...

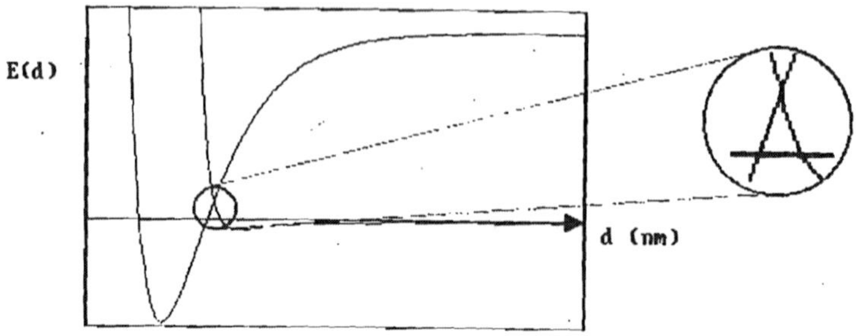

Fig. 4.12.

there is a substantial barrier to chemisorption. Such a barrier has a major influence on the *Kinetics* of adsorption.

Note - The depth of the physisorption well for the hydrogen molecule is actually very small (in some cases negligible), but this is not the case for other molecules and does not alter the basic conclusions regarding dissociative adsorption that result from this model; namely that the process may be either activated or non-activated depending on the exact location of the curve crossing.

At this point it is useful to return to consider the effect of such a barrier on the relationship between the activation energies for adsorption and desorption, and the energy (or enthalpy) of adsorption.

Clearly, from the diagram

$$E_a^{des} - E_a^{ads} = -\Delta H_{ads}$$

but, since the activation energy for adsorption is nearly always very much smaller than that for desorption, and the difference between the energy and enthalpy of adsorption is also very small, it is still quite common to see the relationship.

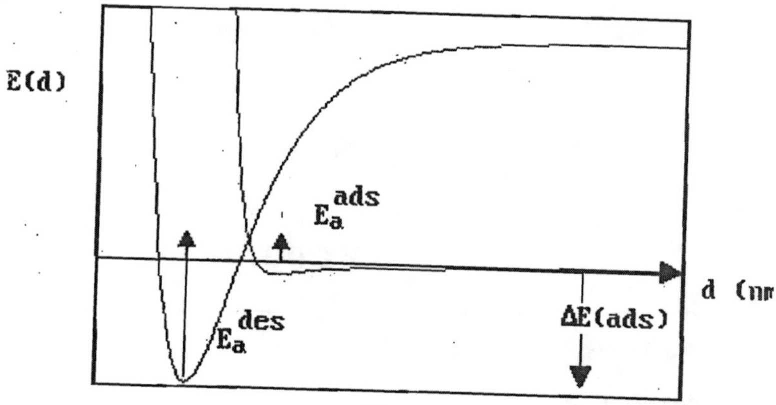

Fig. 4.13

$$E_a^{des} \sim -\Delta H_{ads}$$

For a slightly more detailed treatment of the adsorption process, you are referred to the following examples of More Complex PE Curves & Multi-Dimensional PE Surfaces.

5

Desorption

Introduction

A process in which atomic and molecular species residing on the surface of a solid leave the surface and enter the surrounding gas or vacuum. In stimulated desorption studies, species residing on a surface are made to desorb by incident electrons or photons. Measurements of these species provide insight into the ways that radiation affects matter, and are useful analytical probes of surface physics and chemistry. In thermal desorption studies, adsorbed surface species are caused to desorb as the sample is heated under controlled conditions. These measurements can provide information on surface-bond energies, the species present on the surface and their coverage, the order of the desorption process, and the number of bonding states or sites.

Stimulated desorption

Stimulated desorption from surfaces is initiated by electronic excitation of the surface bond by incident electrons or photons. The classical model of desorption is an adaptation of the theory of gas-phase dissociation, in which desorption results from excitation from a bonding state to an antibonding state. In this model the excitations for gas-phase dissociation and desorption from surfaces are identical, but the surface, which can absorb energy from the excited adsorbate before it can desorb, dramatically reduces the overall dissociation yield from surfaces. Thus, processes which are strong in the gas phase can be quenched on the surface, leading to a marked difference in the relative importance of competing mechanisms in the two environments. This model applies predominantly to neutral desorption of adsorbed molecular species.

Another model, proposed by M.L. Knotek and P.J. Feibelman, which is more applicable to the phenomenon of ion desorption, was first observed in studies of the desorption of positively ionized oxygen (O^+) from the surface of titanium (IV) oxide (TiO_2). Here it is found that O^+ is desorbed not by valence level excitation, but by ionization of the titanium and oxygen core levels. These levels, of course, have little to do with bonding. Furthermore, the fact that the oxygen is desorbed as an O^+ ion (whereas it is nominally at O^{2-} on the surface) implies a large (three-electron) charge-transfer preceding desorption.

Titanium (IV) oxide is an ionic solid and more importantly it is a maximal a valency ionic compound. Maximal valency means that in the solid the titanium (Ti) is ionised to the noble-gas configuration Ti^{4+}, so that there are effectively no valence electrons on the Ti ion in the solid; its highest occupied orbital is 3p at about 30 eV below the Fermi level. The important thing about maximal valency ionic materials is that if a core level is inonized in either the cation or the anion, higher-lying valence electrons will decay into the deep hole left in the core level and give up the energy gained by emitting multiple-valence electrons from the solid (multiple Auger emission). Since valence electrons are only on the anion, the anion is selectively stripped in the process. In TiO_2 one electron is lost from O^{2-} by decay into the core hole; in about 10% of the events two Auger electrons are emitted, transforming the O^{2-} to an O^+ ion. An O^+ ion in a highly repulsive potential for positive ions results, which leads to O^+ desorption. In the general case of an ionic bond M^+A^-, the incident radiation ionizes the core level of the M^+A^- pair. Auger decay of the core hole selectively removes charge from the A^- ion and results in its being transformed into an A^+, and the resulting repulsion causes the A^+ to desorb. This model applies to a wide range of ionic materials, and explains why maximal valency materials like TiO_2, vanadium (V) oxide (V_2O_5), tungsten (VI) oxide (WO_3), molybdenum (VI) oxide (MoO_3), sodium fluoride (NaF), lithium fluoride (LiF), sodium chloride (NaCl), an so forth, are decomposed by irradiation. See also: *Auger effect*.

This mechanism for desorption can also be effective for covalently bonded surface species. Covalently bonded structures differ from ionically bonded structures in that the electronic charge which bonds the atoms together is distributed more evenly through the structure. The electronic interaction between atoms is strong so that an electronic excitation produced on one atom can quickly move to a neighbour. Thus, a simple electronic excitation can quickly move off the site where it was created before dissociation can occur. It typically takes 10^{-12} s for bond dissociation to occur, whereas the motion of simple electronic excitations can occur in 10^{-16}s. However, an important attribute of states where two or more electrons are removed from the atom is that such multiply excited states are much more difficult to move to a neighbour and hence are localised for times long enough for desorption to occur. Several features of a covalent bond can enhance this tendency to localise on the atom—for example, reducing the number of neighbours (and hence bonds), or reducing the bond's ability to transport charge to the desorbing species to neutralise the excited state.

Applications

Stimulated desorption studies are finding wide use. First, they can show the ways in which radiation affects the structure of solids. This will have important applications in the areas of radiation-induced damage and chemistry. Second, as an analytical tool, they offer a unique new way to study the physics and chemistry of atoms on surfaces which, when combined with the many other surface techniques based largely on electron spectroscopy, can provide new insight. Finally, models of the surface bond are put to a much sterner test in attempting to explain desorpton phenomena. See also: *Radiation damage to materials*.

Analysis of surface structure

The techniques of electron - and photon - stimulated desorption can be very useful in the study of how atoms and molecules bond to surfaces. Since ions are desorbed only from the topmost layers of the solid, these techniques study only that outermost layer. By mass-analyzing the desorbed atomic and molecular species, the nature of the chemical species on the surface can be deduced. More importantly, an adsorbed species will be desorbed when the core level of its bonding-site atom is ionised. Since the energy of the electron or photon necessary to ionize a given atom's core level is well known and characteristic of that atom, the bonding sites of each of the desorbing species can be determined. Thus surface-, site-, and adsorbate-specific information can be obtained. Elaborate methods for deducing how the bonding site and adsorbate atoms are electronically and structurally configured can be utilized with this technique. The method is almost the only spectroscopy capable of detecting hydrogen, a classic problem in this field.

Ion angular distributions

An additional important discovery is that ion angular distributions from stimulated desorption are not isotropic, but show that ions are emitted in relatively narrow cones which project along the nominal ground-state bond directions. An angular display shows an azimuthal pattern which reflects the symmetry of the bonding site, and a polar angular distribution which reflects the bonding angle relative to the surface normal. Thus this technique provides a direct display of the surface-bonding geometry.

There are a number of ways to measure the angular distributions of desorbed ions. One of the simplest employs an image intensifier. Individual ion events result in light pulses at a phosphor screen, and time-averaged photographs of the screen provide visual images of the distributions and thus allow a direct view of the arrangement of atoms and molecules on the surface. The directness and simplicity of the technique make it particularly attractive to measure complicated surface adsorbate structures or chemical configurations which do not contain the long-range order necessary in many other techniques for measuring structure. *See also light amplifier.*

Thermal desorption

Thermal desorption mass spectroscopy is possibly the

oldest technique for the study of adsorbates on surfaces. Three primary forms of the thermal desorption experiment involve measurement of (1) the rate of desorption from a surface during controlled heating (temperature-programmed thermal desorption), (2) the rate of desorption at constant temperature (isothermal desorption), and (3) surface lifetimes and diffusion under exposure to a pulsed beam of adsorbates (molecular-beam experiments). Of the three, temperature-programmed thermal desorption is by far the most widely applied, and the discussion below is confined to it. See also: *Molecular beams.*

In thermal desorption measurements, a surface is first dosed with an adsorbate gas, and then the sample is heated at a controlled rate β while the rate of desorption of the adsorbed gas is measured with a mass-filter gas analyzer. The most straightforward information provided is the nature of the adsorbed species from mass analysis, and a determination of the absolute coverage by the adsorbate, which is very difficult to obtain with other techniques. Coverage is determined by measuring the total amount of gas desorbed. The technique can also provide several important kinetic parameters of the desorption process as follows.

Since adsorption reactions are usually non activated, the desorption activation energy *E*, which thermal desorption can provide, is a good approximation to the differential heat of adsorption. Thus these studies provide information on surface-bond energies. See also: *Adsorption.*

Desorption kinetics can be quite complex, especially when there are strong interactions between adsorbates. The general form of the *m*th-order desorption rate equation is given by Eq. (1), where R(t) is

$$R(t) = V_0^{(m)} n^m \exp(-E/kTs) \qquad \ldots (1)$$

the rate of desorption as a function of time, T_s is the substrate temperature, $\upsilon_0^{(m)}$ is the preexponential factor of the desorption-rate coefficient for order *m*, and *n* is the two-dimensional adsorbate concentration. Desorption has been observed to obey zero-, first-, and second-order kinetics. Zero-order kinetics is observed for multilayers where removal of an adsorbed species does not lower the effective coverage feeding the desorption process. First-order kinetics is observed for simple desorption of an atomic or molecular species from a submonolayer of adsorbate. Second-order kinetics is often observed in systems where dissociative adsorption has occurred and recombination reactions occur before the desorption.

Desorption spectra can determine the number of binding states of an adsorbate on a surface, from the number of desorption peaks observed, and, coupled with structural models of the surface, can help determine both the nature of the bonding sites and the bonding geometries.

The preexponential factor of the desorption-rate coefficient, equivalent to an attempt frequency, the rate at which the desorbing particle attempts to leave the surface, not only is necessary to understand desorption kinetics but helps in understanding the general problem of surface chemical dynamics.

The existence of indirect and direct interactions between coadsorbed species can be determined from thermal desorption. Knowledge of such interactions is directly applicable to such important problems as catalysis and corrosion. See also *Catalysis; Corrosion*.

There are several methods to derive the kinetic parameters of Eq. (1) experimentally. If a value is assumed for the preexponential factor$_{vo}$ (usually of the order of 10^{13}), the desorption activation energy is given by Eq. (2) for first-order kinetics, and by Eq. (3)

$$E/kT_p = (v_0^{(1)}/\beta) \exp(-E/kT_p) \qquad \text{... (2)}$$

for second-order kinetics. Here β is the heating

$$E_k T_p^{\,2} = (v_0^{(2)} n_o/\beta) \exp(-E/kT_p) \qquad \text{... (3)}$$

rate, n_0 is the initial adsorbate concentration, and T_p is the temperature of peak desorption rate. Alternatively, β can be varied and *E* can be derived from Eq. (4) for first order; for second order, E/k is determined

$$E/kT_p = d(\ln\beta)/d(\ln T_p) \qquad \text{... (4)}$$

from the slope of a plot of In $(n_0 T^2_p)$ versus $1/T_p$. It is also possible to derive the desorption energy from the width of the desorption peak.

While the thermal desorption techniques are among the simplest of surface probes, they remain indispensable because of their directness and the variety of information they convey. Thus while surface science moves to detailed methods involving extremely sophisticated apparatus, the simple thermal desorption methods remain an important part of the overall picture. See also *Surface physics*

Desorption Process

An adsorbed species present on a surface at low temperatures may remain almost indefinitely in that state. As the temperature of the substrate is increased, however, there will come a point at which the thermal energy of the adsorbed species is such that one of several things may occur:

1. a molecular species may decompose to yield either gas phase products or other surface species.
2. an atomic adsorbate may react with the substrate to yield a specific surface compound, or diffuse into the bulk of the underlying solid.
3. the species may desorb from the surface and return into the gas phase.

The last of these options is the desorption process. In the absence of decomposition, the desorbing species will generally be the same as that originally adsorbed but this is not necessarily always the case.

(An example where it is not is found in the adsorption of some alkali metals on metallic substrates exhibiting a high work function where, at low coverages, the desorbing species is the alkali metal ion as opposed to the neutral atom. Other examples would include certain isomerisation reactions.)

Desorption Kinetics

The rate of desorption of an adsorbate from a surface can be expressed in the general from:

$R_{des} = k\, N^x$ where

x - kinetic order of desorption

k - rate constant for the desorption process

N - surface concentration of adsorbed species.

The order of desorption can usually be predicted because we are concerned with an elementary step of a "reaction": specifically,

I. Atomic or simple Molecular Desorption

$$A_{(ads)} \rightarrow A_{(g)}$$
$$M_{(ads)} \rightarrow M_{(g)}$$

— will usually be a first order process (i.e. x = 1). Examples include...

$W / Cu_{(ads)} \rightarrow W_{(s)} + Cu_{(g)}$; desorption of Cu atoms from a W surface

$Cu / CO_{(ads)} \rightarrow Cu_{(s)} + CO_{(g)}$; desorption of CO molecules from a Cu surface

II. Recombinative Molecular Desorption

$$2\,A_{(ads)} \rightarrow A_{2\,(g)}$$

— will usually be a second order process (i.e. x = 2). Examples include...

$Pt / O_{(ads)} \rightarrow Pt_{(s)} + O_{2\,(g)}$; desorption of O atoms as O_2 from a Pt surface

$Ni / H_{(ads)} — Ni_{(s)} + H_{2(g)}$; desorption of H atoms as H_2 from a Ni surface

The rate constant for the desorption process may be expressed in an Arrhenius form,

$$k_{des} = A \exp(-E_a^{des} / RT)$$

where E_a^{des} is the activation energy for desorption, and

A is the pre-exponential factor; this can also be considered to be the "attempt frequency", ν, at overcoming the barrier to desorption.

This then gives the following general expression for the rate of desorption

$$R_{des} = -\,dN/_{dt} = v.\ N^{x}\ .\exp\left(-E_a^{des}/_{RT}\right)$$

In the particular case of simple molecular adsorption, the pre-exponential/frequency factor (v) may also be equated with the frequency of vibration of the bond between the molecule and substrate; this is because every time this bond is stretched during the course of a vibrational cycle can be considered an attempt to break the bond and hence an attempt at desorption.

Surface Residence Times

One property of an adsorbed molecule that is intimately related to the desorption kinetics is the *surface residence time* - this is the average time that a molecule will spend on the surface under a given set of conditions (in particular, for a specified surface temperature) before it desorbs into the gas phase.

For a first order process such as the desorption step of a molecularly adsorbed species:

$$M_{(ads)} - M_{(g)}$$

the average time (τ) prior to the process occurring is given by:

$$\tau = 1/k_1$$

where k_1 is the first order rate constant (no proof given).

From the previously derived desorption formulae we know that

$$k_1 = v\ \exp\ (-E_a^{des}\ /RT)$$

and if we also substitute for E_a^{des} using the approximate relation $E_a^{des} \sim -\Delta H_{ads}$.

then we get the following expression for the surface residence time

$$\tau = \tau_o\ \exp\ (-\Delta H_{ads}\ /\ RT)$$

where: τ_0 (= 1/v) corresponds to the period of vibration of the bond between the adsorbed molecule and substrate and is frequently taken to be about 10^{-13}s.

6

Surface and Interfacial Chemistry

Introduction

Chemical processes that occur at the phase boundary between gas-liquid, liquid-liquid, liquid-solid, or gas-solid interfaces.

The chemistry and physics at surfaces and interfaces govern a wide variety of technologically significant processes. Chemical reactions for the production of low-molecular-weight hydrocarbons for gasoline by the cracking and reforming of the high-molecular-weight hydrocarbons in oil are catalyzed at acidic oxide materials. Surface and interfacial chemistry are also relevant to adhesion, corrosion control, tribology (friction and wear), microelectronics, and biocompatible materials. In the last case, schemes to reduce bacterial adhesion while enhancing tissue integration (Fig. 6.1) are critical to the implantation of complex prosthetic devices, such as joint replacements and artificial hearts. Selected technologies that involve surfaces and interfaces are listed in Table 6.1. See also: *Biomedical chemical engineering; Cracking; Heterogeneous catalysis; Prosthesis; Surface physics*

TABLE 6.1
Some Typical Surface and Interfacial Processes

Interface	*Processes*	*Significance*
Liquid-liquid	Solute partitioning	Solvent Extraction
Gas-solid	Adsorption/desorption Surface-mediated reactions	Catalysis
Liquid-solid	Electrochemistry Molecular recognition	Solar energy conversion Energy transfer and storage in cell membranes DNA sequencing

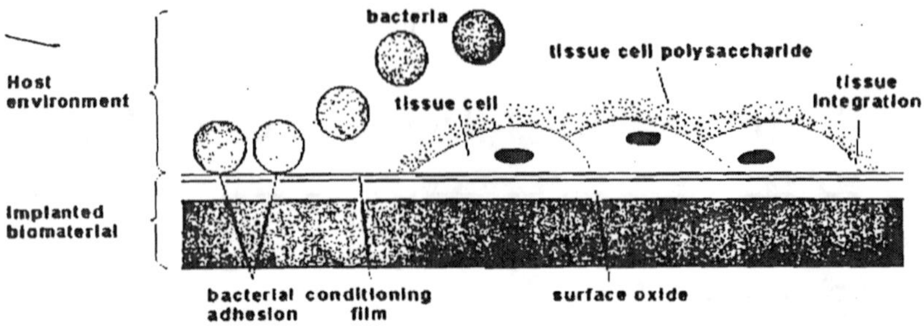

Fig. 6.1: Scheme for reducing bacterial adhesion at an implanted biomaterial.

Interactions with the substrate may alter the electronic structure of an adsorbate. Those interactions that lower the activation energy of a chemical reaction result in a catalytic process. Adsorption of reactants on a surface also confines the

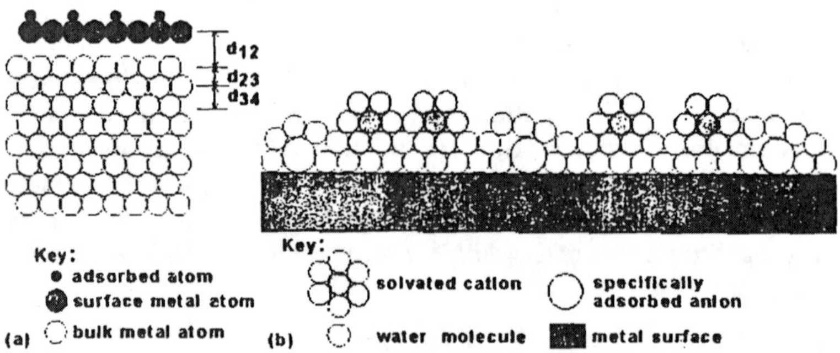

Fig. 6.2: Microscopic surface structures. *(a)* Surface atoms occupying positions other than the bulk-terminated locations. The first interlayer spacing d_{12} is greater than the bulk interlayer spacing d_{34}. The second interlayer spacing d_{23} is smaller than the bulk interlayer spacing. *(b)* Electrical double layer. The plane of closest approach is the inner Helimholtz plane. The adjacent layer of solvated cations is the outer Helmholtz plane.

reaction to two dimensions as opposed to the three dimensions available for a homogeneous process. The two-dimensional confinement of reactants in a bimolecular event seems to drive biochemical processes with higher reaction efficiencies at proteins and lipid membranes. See also: *Adsorption*

A limitation in the study of surfaces and interfaces rests with the low concentrations of the participants in the chemical process. Concentrations of reactants at surfaces are on the order of 10^{-10} to 10^{-8} mole/cm^2. Such low concentrations pose a sensitivity problem from the perspective of surface analysis. Experimental techniques with high sensitivity are required to examine the low concentrations of a surface species at interfaces (Fig. 6.2)

Electron Surface Analysis Techniques

Electron spectroscopy methods are widely used in the study of surfaces because of the small penetration depth of electrons through solids. This attribute makes electron spectroscopy inherently surface-sensitive, since only a few of the outermost atomic layers are accessible. Electron spectroscopy requires propagation over distances of ~10 cm without collisions with gas-phase molecules. The mean free path of an electron, λ_e,

$$\lambda_e = \frac{4kT}{\pi\xi^2 P} \qquad \text{... (1)}$$

through a gas is given by Eq. (1), where k is Boltzmann's constant, T is absolute temperature, ξ is molecular diameter of the ambient gas, and P is ambient pressure. If an electron propagates 10 cm (a typical path length between the sample and detector) at room temperature, a vacuum of 7×10^{-2} pascal is required. Electron spectroscopy is typically carried out at much lower pressures of ultrahigh vacuum ($<1 \times 10^{-7}$ Pa) to ensure sample integrity.

The methods of electron spectroscopy used in surface studies have several common characteristics (Table 6.2). A source provides the incident radiation to the sample, which can be in the form of electrons, x-radiation, or ultraviolet radiation. Electron beams are generated from the thermionic emission of metal filaments or metal oxide pellets. The incident radiation induces an excitation at the surface of the sample, which alters the energy distribution of electrons that leave the surface. This distribution provides a diagnostic of the composition or structure of the interface. Typical electron energy analyzers include retarding-field, cylindrical-mirror, and hemispherical-mirror analyzers. See also: *Electron spectroscopy*

TABLE 6.2
Surface-sensitive Experimental Techniques

Technique	*Source**	*Detectors*	*Level of Information*
Auger electron spectroscopy (AES)	Electrons 2-3 keV	Cylindrical mirror or retarding field	Elemental composition
X-ray photoelectron speotroscopy (XPS)	X-rays	Hemispherical or cylindrical mirror	Elemental composition and oxidation state
	1254 eV (Mg)		
	1487 eV (Al)		
Ultraviolet photoelectron spectrocopy (UPS)	UV radiation	Hemispherical or cylindrical mirror	Electronic properties of adsorbate and/or bulk material
	21 eV He (I)		
	41 eV He (II)		
Energy loss spectroscopy (ELS)	Electrons 50-1000 eV	Electron energy analyzer	Electronic structure of surface
High-resolution electron energy loss spectroscopy (HREELS)	Electrons	Electron energy analyzer	Vibrational losses
	1-10 eV		
Low-energy electron diffracton (LEED)	Electrons 20-500 eV	Retarding fields and phosphorescent screen	Surface structure or periodicity
Infrared spectroscopy (IRS)	Photons	Mercury-cadmium--telluride or indium antimony	Molecular identity
Optical ellipsometry	Photons	Photomultiplier	Adsorbate layer thickness
Scanning tunneling microscopy (STM)	Tunneling current	Ammeter	Substrate roughness and texture

*Mg = magnesium; Al = aluminum; He = helium.

Auger Electron Spectroscopy (AES)

This technique is used to determine the elemental composition of a surface species. As shown in Fig. 6.3a, the incident electron beam (2-3 keV) ionises a K-shell electron of an atom of energy, E_k, creating an electron vacancy. An incident x-ray can also initiate the same event. An electron in a higher-energy L-shell with energy E_L subsequently drops down to fill the vacancy, releasing the energy $E = E_k - E_L$. The excess energy can either be emitted as an x-ray or be released to an outer-shell electron. Known as an Auger electron, it is ejected with kinetic energy $E \sim (E_k - E_L - E_M)$. This is a *KLM* transition, based on the energy levels involved in this particular process. Other transitions involving different electrons also occur (for example, *KLL* and *LMM*). The kinetic energy of the Auger electrons provide a so-called fingerprint for elemental identification, as found in numerous reference compilations. See also: *Auger effect; Electron-hole recombination*

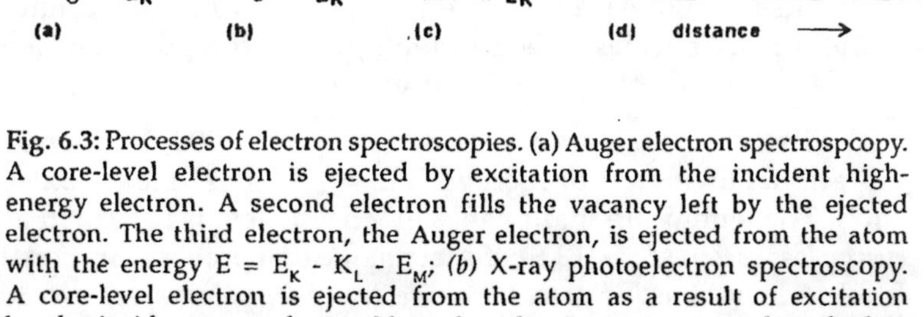

Fig. 6.3: Processes of electron spectroscopies. (a) Auger electron spectrospcopy. A core-level electron is ejected by excitation from the incident high-energy electron. A second electron fills the vacancy left by the ejected electron. The third electron, the Auger electron, is ejected from the atom with the energy $E = E_K - K_L - E_M$; *(b)* X-ray photoelectron spectroscopy. A core-level electron is ejected from the atom as a result of excitation by the incident x-ray beam. Note that the Auger process, described in *a*, can also occur as a result of x-ray stimulation of the sample. *(c)* Ultraviolet photoelectron spectroscopy. Ultraviolet radiation excites the sample, causing ejection of weakly bound valence electrons. *(d)* Electron energy loss spctroscopy. An electron is excited into a higher electronic (ELS) or vibrational (HREELS) state by the incident electron beam. The scattered electrons lose the energy that was required for excitation. E_V = energy of the valence electrons.

X-ray Photoelectron Spectroscopy (XPS)

This technique is used to determine the oxidation state as well as the elemental composition of a surface species. In x-ray photoelectron spectroscopy, an inner-shell electron or photoelectron is ejected as a result of excitation by an incident x-ray photon (Fig. 6.3b). The x-ray source is usually a magnesium or aluminum anode. The ejected electrons have an energy $E = h_v - E_i$, where h_v and E_i are the energies of the x-ray and the inner electron, respectively. The resulting vacancy can be filled by the Auger relaxation process. The energy of the photoelectron reflects both the oxidation state and elemental composition of the surface species. See also: *X-ray spectrometry.*

Ultraviolet Photoelectron Spectroscopy (UPS)

This is similar to x-ray photoelectron spectroscopy except that ultraviolet photons are used for excitation. The helium I and II lines are the most commonly used sources. Since ultraviolet radiation is less energetic than x-radiation, ultraviolet photoelectron spectroscopy probes only the valence electron levels of the surface species. The electron is ejected from the other outer shell with energy $E = h_v - E_u$, where E_u is the energy of the valence electron (Fig. 6.3c)

Electron Energy Loss Spectroscopy (EELS)

This technique probes the electronic and vibrational states of the surface species. It is divided into two categories, based on the energy E_p of the incident electron.

Electronic levels are accessed with E_p = 50-1000 eV, whereas vibrational levels are probed with E_p = 1-10 eV (1 meV = 8.066 cm^{-1}). The technique that probes electronic levels is known as *energy loss spectroscopy* (ELS); the technique that probes vibrational levels is known as *high-resolution electron energy loss spectroscopy* (HREELS). Higher resolution is required to resolve the vibrational bands because the energy transitions are smaller, as shown in Fig. 6.3d. Most of the electrons are scattered elastically, and they fail to induce an electronic or vibrational excitation of the surface species. However, in a few cases the incident electron undergoes an inelastic collision, losing energy via excitation of a vibration or electronic transition. The energy of the reflected beam is $E = E_p - E_{loss}$ (Fig. 6.3d). For high-resolution electron energy loss spectroscopy, only those vibrations that have a component of their transition dipole moment perpendicular to the surface can be excited. With this surface selection rule, the loss spectrum often provides information pertaining to the orientation of the surface species with respect to the surface normal.

Low-Energy Electron Diffraction (LEED)

This technique is the surface analogy of x-ray diffraction; it probes the two-dimensional periodicity of the surface species. Applicable primarily to studies of single-crystal surfaces, low-energy electron diffraction arises from the elastic scattering of electrons incident on the surface. The electrons scatter in phase or out of phase, depending on the incident energy (wavelength) and the position of the atoms relative to one another. In-phase scattering, which interferes constructively to yield the low-energy electron diffraction pattern, occurs when the Bragg condition is satisfied. The Bragg condition is given by Eq. (2), where *a* is the distance

$$a(\sin\phi - \sin\phi_0) = n\lambda \qquad \dots (2)$$

between surface atoms, ϕ and ϕ_0 are the backscattered and incident angles of the electron beam, λ is the wavelength of the electron beam, and *n* is an integer corresponding to the diffraction order. The diffracted electrons are accelerated and focused by a series of grids onto a phosphorescent screen for visualization of the low-energy electron diffraction pattern. See also *Electron diffraction*

Optical Surface Analysis Techniques

Optical spectroscopy techniques (visible and infrared) are also useful for probing the chemical composition and molecular arrangement of surface species. Typical application configurations are the transmission and reflection (both external and internal) modes as shown in Fig.6. 4. Transmission spectroscopy relies on the passage of the probe beam through the sample. External and internal reflection spectroscopies involve the reflection of the probe beam from a medium with a lower refractive index to a medium with a higher refractive index, and from a higher to lower refractive index, respectively. The sample support must be optically transparent to the probe beam for the internal reflection mode. In both cases, the substrates are polished to a smooth, mirrorlike finish. See also *spectroscopy*

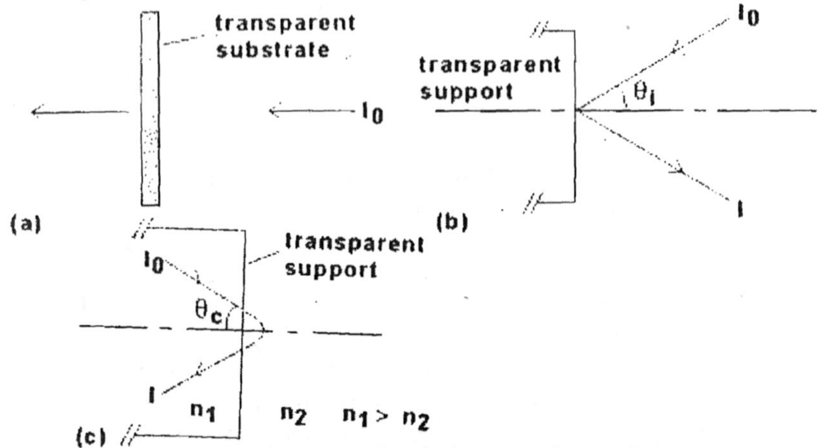

Fig. 6.4: Typical configuration for application of optical spectroscopies to surface analysis. *(a)* Transmission. *(b)* External or specular reflection. *(c)* Internal reflection. I = reflected light intensity. I_0 = initial light intensity. θ_i = angle of incidence at phase 1-phase 2 boundary. θ_c = angle of refraction at phase 1-phase 2 boundary, n_1 = refractive index of phase 1. n_2 = refractive index of phase 2.

Diffuse reflection spectroscopy, which is essentially reflection from a rough surface, has been utilised to examine supported-metal catalysts and high-surface-area oxides. In all cases, the power lost by the interaction of the incident radiation with the surface species gives rise to an absorption spectrum. Comparison with the spectrum of the adsorbate starting material provides insights into the nature of the surface species. Propagation distances are much longer than those for electron spectroscopy, allowing the application of these techniques to in situ surface studies such as at the gas-solid interface of a high-pressure chemical reactor as well as the liquid-solid interface of an electrochemical cell. Raman spectroscopy also provides information about the molecular composition of the surface species, although at a lower sensitivity. See also *Raman effect*

Optical probes are less surface-sensitive than electron spectroscopy because of the longer wavelength of the incident energy. This can result in the sampling of both bulk and surface species, complicating the interpretation of the observed spectrum. Penetration depths for optical spectroscopies are typically a fraction of the wavelength of the incident light, and they are a function of the optical properties of the sample and of the support and the angle of incidence. For visible radiation, penetration depths in a reflection experiment are tens to hundreds of nanometers, whereas those for infrared radiation are a few hundred to thousands of nanometers. For internal reflection spectroscopy, the penetration depth d_p is given by Eq. (3),

$$d_p = \frac{\lambda_1}{2\pi\ (\sin^2\ \theta - n_{21})^{1/2}} \quad \ldots (3)$$

where λ_1 is the wavelength of the incident radiation, θ is the angle of incidence, and n_{21} is the ratio of the refractive indices of media 2 and 1, respectively.

Infrared Spectroscopy (IRS)

This is a useful technique for the determination of the molecular composition and orientation of the surface species. In an external reflection mode, a beam of infrared light is reflected at a smooth metal, semiconductor, or insulator surface, as shown in Fig. 6.4b. For films with thicknesses greater than ~ 10 nm, conventional dispersive spectrometers can be used. However, studies of monolayers require a Fourier transform spectrometer and high-sensitivity, low-noise, photo conductive detectors, for example, mercury-cadmium-telluride.

The band shapes and intensities of reflection spectra can be calculated quantitatively with classical electromagnetic theory. Variables in the calculation, which is a reflection spectroscopy analog to the Beer-Lambert law in transmission spectroscopy, include the optical function of each phase and the polarization and angle of incidence of the incident beam. Comparison of the intensities of the observed and calculated spectra, assuming that the force constants of a given vibrational mode are unchanged by immobilisation, provides an estimate of the average spatial orientation of the adsorbate with respect to the substrate. As with high-resolution electron energy loss spectroscopy, such interpretations at highly reflective surfaces such as metals are aided by the preferential excitation of those vibrational modes with a transition dipole moment perpendicular to the surface, according to the infrared surface selection rule. An orientational analysis at a surface with a low reflectivity can also be accomplished, but it is hindered by a complex variation of the relative degree of excitation of vibrations with modes both perpendicular and parallel to the surface. See also: *Infrared spectroscopy*

Optical Ellipsometry

This is a variation of the external reflection measurement in that the state of polarization of the reflected light is measured as opposed to the reflected power. The change in polarization is caused by the difference in the intensity and phase change for the components of the incident electromagnetic wave that are parallel and perpendicular to the surface. Such changes, which are a function of the optical properties of the adsorbate, can be used to determine its thickness. More detailed treatments with effective medium theory can also provide insights into the density of the surface layer. Computerized data acquisition is advantageous, because numerical analysis requires the collection of large data sets and extensive complex algebraic manipulation. See also: *Ellipromettry*

Scanning Tunneling Microscopy (STM)

This is a surface analysis technique that provides information about the three-dimensional topography of a surface with resolution at the atomic level. Operation in ultrahigh vacuum, at liquid-solid interfaces, and in laboratory ambient (air) has been demonstrated.

The two general operation configurations are the constant-current and constant-height modes. In the constant-current mode (Fig. 6.5a), a bias voltage (~10-10,000 mV) is applied between a sharp metallic tip and a conducting sample. A small tunneling current (~0.2-10 nanoamperes) flows when the tip is positioned within a few atomic diameters of the sample. The tip is then scanned laterally with a piezoelectric translator over a surface while a computer-controlled feedback mechanism adjusts the height of the tip to maintain a constant current. An image of the surface contour is constructed by plotting a weighted average of the tip height z versus the lateral positions x and y.

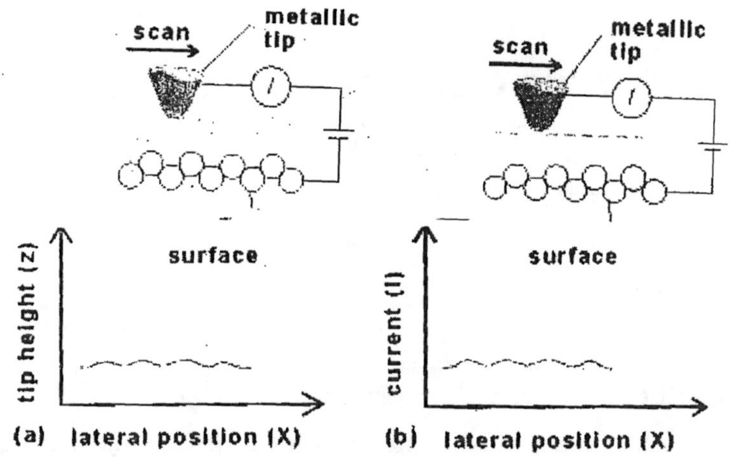

Fig. 6.5: Scanning tunneling microscopy and images for *(a)* constant-current mode and *(b)* constant-height mode; shown only in x dimension for simplicity.

In the constant-height mode, a tip is scanned across a surface at constant height while monitoring the current I (Fig. 6.5b). A surface image is derived from a plot of multiple scans displaced from each other.

Each mode of operation has its own specific advantages. The constant-current mode provides topographic pictures of surfaces that are not atomically flat. This mode operates at scan rates of only a few hertz and is limited by the vertical movement of the tip by the z translator. In contrast, the constant-height mode can image atomically smooth surfaces at rates up to ~ 1 kHz. These high imaging rates are attainable because only the electronics for monitoring the tunneling current—and not the vertical movement of the z translator —must respond to the atoms passing under the tip. A high imaging rate allows the examination of real-time processes, reduces data acquisition time, and minimizes image distortion caused by thermal drift and piezoelectric creep.

7

Adsorbate Geometries and Structures on Metals

Introduction

We can address the question of what happens when a molecule becomes adsorbed onto a surface at two levels; specifically we can aim to identify.

1. the nature of the adsorbed species and its local adsorption geometry (i.e. its chemical structure and co-ordination to adjacent substrate atoms).
2. the overall structure of the extended adsorbate/ substrate interface (i.e. the long range ordering of the surface).

The latter topic is covered in *Classification of Over layer Structures* whist this chapter will consider only the local adsorption geometry and adsorbate structure.

Chemisorption, by definition, involves the formation of new chemical bonds between the adsorbed species and the surface atoms of the substrate - basically the same type of bonds that are present in any molecular complex. In considering what type of species may be formed on a metal surface, therefore, it is important not to abandon chemical common sense and, if in doubt, to look for inspiration at the structures of known *metal-organic complexes.*

Chemisorption of Hydrogen and Halogens

Hydrogen (H_2)

In the H_2 molecule, the valence electrons are all involved in the H-H σ-bond and there are no additional electrons which may interact with the substrate atoms. Consequently, chemisorption of hydrogen on metals is almost invariably a dissociative process in which the H-H bond is broken, thereby permitting the hydrogen atoms to independently interact with the substance The adsorbed species in this instance therefore are hydrogen atoms.

The exact nature of the adsorbed hydrogen atom complex is generally difficult to determine experimentally, and the very small size of the hydrogen atom does mean that migration of hydrogen from the interface into sub-surface layers of the substrate can occur with relative ease on some metals (e.g. Pd, rare earth metals).

The possibility of molecular H_2 chemisorption at low temperatures cannot be entirely excluded, however, as demonstrated by the discovery of molecular hydrogen transition metal compounds, such as $W\,(\eta^2\text{-}H_2)(CO)_3\,(Pi\text{-}Pr_3)_3$, in which both atoms of the hydrogen molecule are coordinated to a single metal centre.

Halogens (F_2, Cl_2, Br_2 etc.)

Halogens also chemisorb in a dissociative fashion to give adsorbed halogen atoms. The reasons for this are fairly clear - in principle a halogen molecule could act as a Lewis base and bind to the surface without breakage of the X-X bond, in practice the lone pairs are strongly held by the highly electronegative halogen atom so any such interaction would be very weak and the thermodynamics lie very heavily in favour of dissociative adsorption [i.e. $D(X\text{-}X) + D\,(M\text{-}X_2) << 2\,D(M\text{-}X^{(-)})$]. Clearly the kinetic barrier to dissociation must also be low or non-existent for the dissociative adsorption to occur readily.

Another way of looking at the interaction of a halogen molecule with a metal surface is as follows: the significant difference in electronegativity between a typical metal and halogen is such that substantial electron transfer from the metal to halogen is favoured. If a halogen molecule is interacting with a metal surface then this transferred electron density will enter the σ* antibonding orbital of the molecule, thereby weakening the X-X bond. At the same time, the build-up of negative charge on the halogen atoms enhances the strength of the metal-halogen interaction. The net result of these two effects when taken to their limit is that the halogen molecule dissociates and the halogen atoms interact with the metal with a strong ionic contribution to the bonding.

Halogen atoms tend to occupy high co-ordination sites on the surface - for example, the 3-fold hollow site on *f*cc (111) surfaces (A) and the 4-fold hollow site on *f*cc (100) surfaces (B).

(A) *Plan View* (B)

This behaviour is typical of atomic adsorbates which almost invariably endeavour to maximise their co-ordination and hence prefer to occupy the highest-available co-ordination site on the surface.

As a result of the electron transfer from the metal substrate to the halogen atoms, each adsorbed atom is associated with a significant surface dipole.

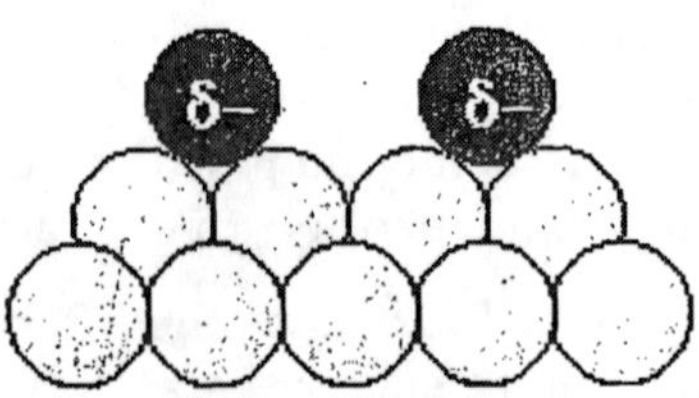

Cross-Section

One consequence of this is that there are repulsive (dipole-dipole) interactions between the adsorbed atoms, which are especially evident at higher surface converges and which can lead to a substantial reduction in the enthalpy of adsorption at specific coverages (if these coverages mark a watershed, above which the atoms are forced to occupy sites which are much closer together).

Another feature of the halogen adsorption chemistry of some metals is the transition from an adsorbed surface layer to surface compound formation at high gas exposures.

Chemisorption of Nitrogen and Oxygen

Oxygen

Oxygen is an example of a molecular which usually adsorb dissociatively, but are also found to adsorb molecularly on some metals (e.g. Ag, Pt). In those cases where both types of adsorption are observed it is the dissociative process that corresponds to the higher adsorption enthalpy.

As noted above, in the molecular adsorption state the interaction between the molecule and the surface is relatively weak. Molecules aligned such that the internuclear axis is parallel to the surface plane may bond to a single metal atom of the surface via both

1. σ-donor interaction, in which the charge transfer is from the occupied molecular π-bonding molecular orbital of the molecule into vacant orbitals of σ-symmetry on the metal (i.e. $M \leftarrow O_2$), and
2. π-acceptor interaction, in which an occupied metal d-orbital of the correct symmetry overlaps with empty p* orbitals of the molecule and the charge transfer is from the surface to the molecule (i.e. $M \rightarrow O_2$).

Although the interaction of the molecule with the surface is generally weak, one might expect that there might be a substantial barrier to dissociation due to the higher strength (and high dissociation enthalpy) of the O=O bond. Nevertheless on most metal surfaces, dissociation of oxygen is observed to be facile which is related to the manner in which the interaction with the surface can mitigate the high intrinsic bond energy and thereby facilitate dissociation.

Once formed, oxygen atoms are strongly bound to the surface and, as noted previously, will tend to occupy the highest available co-ordination site. The strength of the interaction between adsorbate and substrate is such that the adjacent metal atoms are often seen to undergo significant displacements from the equilibrium positions that they occupy on the clean metal surface. This displacement may simply lead to a distortion of the substrate surface in the immediate vicinity of the adsorbed atom (so that, for example, the adjacent metal atoms are drawn in towards the oxygen and the metal-oxygen bond distance is reduced) or to a more extended surface reconstruction.

Dissociative oxygen adsorption is frequently reversible - rather than simply leading to desorption, then heating of an adsorbed oxygen overlayer often results in either the gradual removal of oxygen from the surface by diffusion into the bulk of the substrate (e.g. Si(111) or Cu (111) or to the formation of a surface oxide compound. Even at ambient temperatures, extended oxygen exposure often leads to the nucleation of a surface oxide. Depending on the reactivity of the metal concerned, further exposure at low temperatures may result either in a progressive conversion of the bulk material to oxide or the oxidation process may effectively stop after the formation of a passivating surface oxide film of a specific thickness (e.g. Al).

Nitrogen

The interaction of nitrogen with metal surfaces shows many of the same characteristics as those described above for oxygen. However, in general, N_2 is less susceptible to dissociation as a result of the lower M-N bond strength and the substantial kinetic barrier associated with breaking the $N \equiv N$ triple bond.

Chemisorption of Carbon Monoxide

Depending upon the metal surface, carbon monoxide may adsorb either in a molecular form or in a dissociative fashion - in some cases both states coexist on particular surface planes and over specific ranges of temperature.

1. On the reactive surfaces of metals from the left-

hand side of the periodic table (e.g. Na, Ca, Ti, rare earth metals) the adsorption is almost invariably dissociative, leading to the formation of adsorbed carbon and oxygen atoms (and thereafter to the formation of surface oxide and oxycarbide compounds).

2. By contrast, on surfaces of the metals from the right hand side of the d-block (e.g. Cu, Ag) the interaction is predominantly molecular; the strength of interaction between the CO molecule and the metal is also much weaker, so the M-CO bond may be readily broken and the CO desorbed from the surface by raising the surface temperature without inducing any dissociation of the molecule.

3. For the majority of the transition metals, however, the nature of the adsorption (dissociative v.'s molecular) is very sensitive to the surface temperature and surface structure (e.g. the Miller index plane, and the presence of any lower co-ordination sites such as step sites and defects).

Molecularly chemisorbed CO has been found to bond in various ways to single crystal metal surfaces - analogous to its behaviour in isolated metal carbonyl complexes.

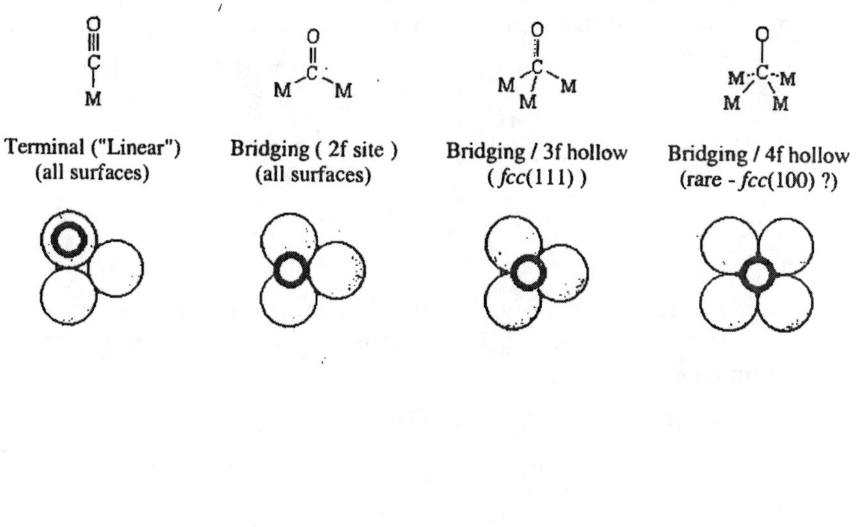

Whilst the above structural diagrams amply demonstrate the inadequacies of a simple valence bond description of the bonding of molecules to surface, they do to an extent also illustrate one of its features and strengths - namely that a given element, in this case carbon, tends to have a specific valence. Consequently, as the number of metal atoms to which the carbon is co-ordinated increases, so there is a corresponding reduction in the C-O bond order.

However, it must be emphasized that a molecule such as CO does not necessarily prefer to bind at the highest available co-ordination site. So, for example, the fact that there are 3-fold hollow sites on an *fcc* (111) surface does not mean that CO will necessarily adopt this site - the preferred site may still be a terminal or 2-fold bridging site, and the site or site(s) which is (are) occupied may change with either surface coverage or temperature. The energy difference between the various adsorption sites available for molecular CO chemisorption appears therefore to be very small.

Chemisorption of Ammonia and other Group V/VI Hydrides

Ammonia has lone pairs available for bonding on the central nitrogen atom and may bond without dissociation to a single metal atom of a surface, acting as a Lewis base, to give a pseudo-tetrahedral co-ordination for the nitrogen atom.

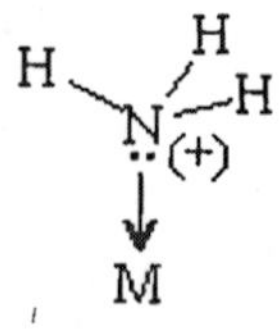

Alternatively, progressive dehydrogenation may occur to give surface NH_x ($x = 2,1,0$) species and adsorbed hydrogen atoms, i.e.

$$NH_3 \rightarrow NH_{2(ads)} + H_{(ads)} \rightarrow NH_{(ads)} + 2H_{(ads)} \rightarrow N_{(ads)}\ 3H_{(ads)}$$

As the numebr of hydrogens bonded to the nitrogen atom

is reduced, the adsorbed species will tend to move into a higher co-ordinaiton site on the surface (thereby tending to maintain the valence of nitrogen).

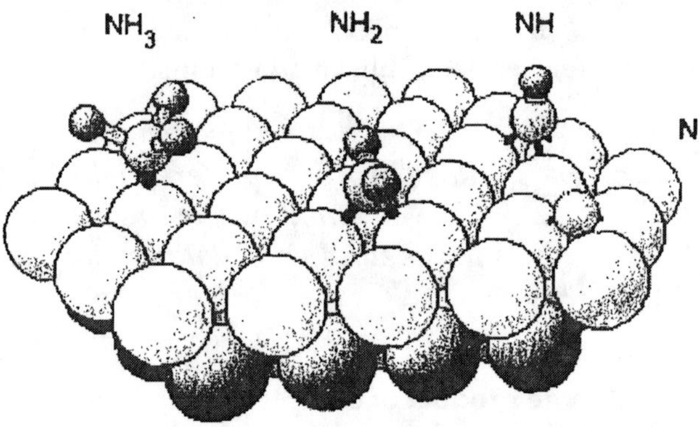

Decomposition fragments of ammonia on an fcc (111) surface

(Picture adapted from the *BALSAC Picture Gallery* by K. Hermann, Fritz-Haber-Institut, Berlin)

Other Group V and Group VI hydrides (e.g. PH_3, H_2O, H_2S) exhibit similar adsorption characteristics to ammonia.

Chemisorption of Unsaturated Hydrocarbons

Unsaturated hydrocarbons (alkenes, alkynes, aromatic molecules etc.) all tend to interact fairly strongly with metal atom surfaces. At low temperatures (and on less reactive metal surfaces) the adsorption may be molecular, albeit perhaps with some distortion of bond angles around the carbon atom.

Ethene, for example, may bond to give both a π-complex (A) or a di-σ adsorption complex (B):

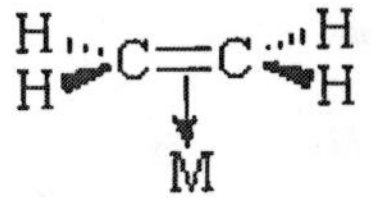

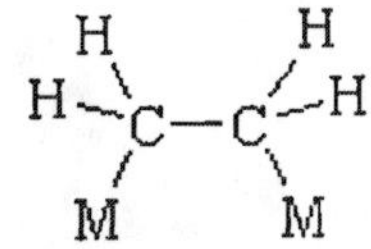

(A) *Chemisorbed* **(B)**
Ethene

[Further examples: *Models of chemisorbed ethene and ethyne on Cu(111)*]

As the temperature is raised, or even at low temperatures on more reactive surfaces (in particular those that bind hydrogen strongly), a stepwise dehydrogenation may occur. One particularly stable surface intermediate found in the dehydrogenation of ethene is the *ethylidyne* complex, whose formation also involves H-atom transfer between the carbon atoms.

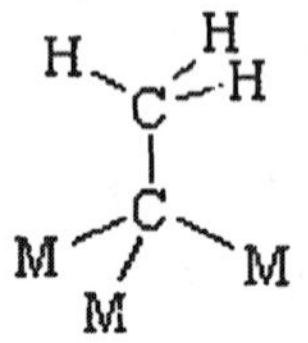

Ethylidyne : this adsorbate preferentially occupies a 3-fold hollow site to give pseudo-tetrahedral co-ordination for the carbon atom.

The ultimate product of complete dehydrogenation, and the loss of molecular hydrogen by desorption, is usually either carbidic or graphitic surface carbon.

8

Adsorption Isotherms

Introduction

The quantity of gas adsorbed by a given amount of solid may be measured at constant temperature (isotherm studies), at constant pressure (isobar studies), or at constant volume (isochore studies). Figure 8.1 shows an experimental arrangement for the determination of isotherms.

Adsorption isotherms have been determined for a large variety of systems at different temperatures and under different conditions of surface preparation. The surfaces studied include different faces of single crystals, thin films, and porous solids with varying distributions of pores. With the exception of a few, all the experimental isotherms may be classified into five types (according to S. Brunauer) as illustrated in Fig. 8.2. Type I, called the Langmuir isotherm, is indicative of monolayer adsorption. The flat part of the isotherm corresponds to saturation of the surface. Chemisorption, unless complicated by further physical adsorption on the chemisorbed surface, is a monolayer process and thus follows the Langmuir isotherm. Type II is indicative of multilayer physical adsorption. The rapid initial rise corresponds to formation of the first adsorbed layer. Type III is observed only in a few physical adsorption studies. Adsorption by porous materials produces isotherms of types IV and V. The desorption curves for these isotherms differ from the adsorption curves (hysteresis) if the pore dimensions are on the order of 10 nm. The explanation for this lies in the

Kelvin equation, according to which there exists a pressure gradient across a curved surface. Hence the pressure at which gases condense into a pore is different than the pressure at which they desorb from the pore.

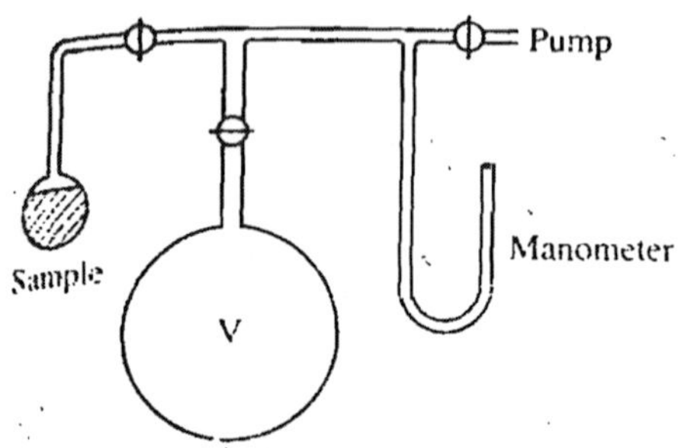

Figure 8.1: Simple apparatus for determining isotherms. The procedure is as follows. Pressure falls when the gas in the vessel V is allowed to expand into the rest of the system. The fall in pressure with the sample is greater than without the sample. From these two pressures and the ideal gas law, one can estimate the number of molecules adsorbed by the sample.

The langmuir Isotherm

In 1918 I. Langmuir demonstrated a theoretical relation between the amount of adsorbate on the surface and its pressure in the gas phase by assuming that the solid substrate provides a certain number of sites for adsorption. The two major premises of the theory are: (1) All the adsorption sites are equivalent, and (2) only one molecule may be adsorbed at each site. From the first premise it follows that adsorption at any given site is not affected by adsorption at neighboring sites. This theory applies only for monolayer coverage since the sites deeper in the solid obviously cannot be equivalent to the surface sites.

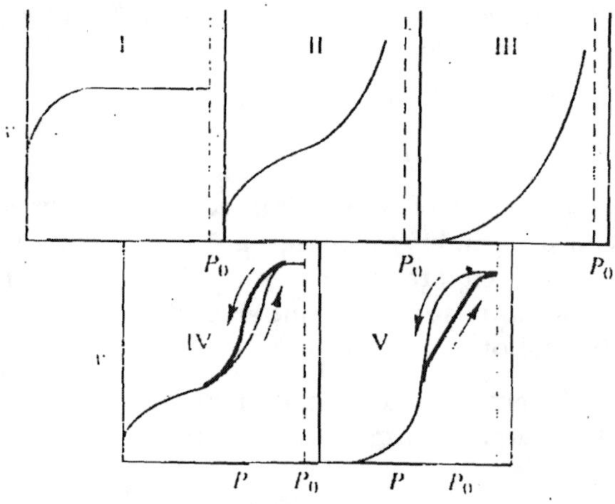

Figure 8.2: Brunauer classification of different isotherms. I: Characteristic of monolayer coverage. See the Langmuir isotherm. II: Most commonly observed type; corresponds to multilayer coverage. III: Infrequently observed; corresponds to multilayer absorption in which K for the first layer (see text) is smaller than K' for subsequent layers. IV and V: Analogous to II and III, respectively, but on porous surfaces with small cavities. Because the pressure across a curved surface may not be the same at equilibrium absorption and desorption pressures are different.

Empirical Isotherms

The Langmuir isotherm is an example of an equation derived from a model (the mechanistic approach). The alternative approach is to find the mathematical function that depicts the observed data and then explore the possible physical basis for such a fit. This is the empirical approach which led to many useful equations of state for gases.

One of the commonly used empirical isoterms is due to H.M.F. Freundlich.

Frendlich Isotherm. It is of the following form.

$$\theta = kP^{1/n} \qquad \ldots (1)$$

where k and n (usually between 1 and 10) are constants. It the adsorption follows the Freundlich isotherm, the corresponding isosteric enthalpy will show a logarithmic decrease with θ:

$$H = -H_0 \ln\theta \qquad \ldots (2)$$

Equation (1) may be derived from Eq. (2), even though the procedure is not simple.

The Tempkin isotherm has the form

$$\theta = A \ln BP \qquad \dots (3)$$

where A and B are constants. This equation is obeyed by systems whose enthalpy deceases at a constant rate with increased surface coverage. The Tempkin isotherm follows from the Langmuir isotherm, under certain conditions, as shown in the following discussion.

From the thermodynamic relation between the Gibbs free energy and the equilibrium constant, we have

$$K = \exp \frac{-\Delta G^\circ}{RT} K' \exp \frac{-\Delta H^\circ}{RT} \qquad \dots (4)$$

where K′ contains the entropy-dependent factors. Since enthalpy decreases linearly with surface coverage, we may write

$$\Delta H^\circ = \Delta H^\circ_{max} - \alpha\theta \qquad \dots (5)$$

where α, the constant, is a measure of the rate of enthalpy decrease. From Eqs. (4), and (5), we get

$$K = B \exp \frac{\alpha\theta}{RT} \qquad \dots (6)$$

where the constant B is given by

$$B = K' \exp \frac{\Delta H^\circ_{max}}{RT} \qquad \dots (7)$$

Consider the behaviour in the region where $\theta \cong 0.5$. Then the Langmuir isotherm gives

$$K = \frac{1}{P}$$

Substituting this into Eq. (3) and rearranging gives the Tempkin isotherm with

$$A = -\frac{RT}{\alpha} \qquad \dots (8)$$

We see from this derivation that the constant B can be obtained from the enthalpy of adsorption at $\theta = 0.5$, and the constant A from the rate of fall in enthalpy with surface coverage.

Langmuir Isotherm

Whenever a gas is in contact with a solid there will be an equilibrium established between the molecules in the gas phase and the corresponding adsorbed species (molecules or atoms) which are bound to the surface of the solid.

As with all chemical equilibria, the position of equilibrium will depend upon a number of factors:

1. The relative stabilities of the adsorbed and gas phase species involved
2. The temperature of the system (both the gas and surface, although these are normally the same)
3. The pressure of the gas above the surface.

In general, factors (2) and (3) exert opposite effects on the concentration of adsorbed species - that is to say that the *surface coverage* may be increased by raising the gas pressure but will be reduced if the surface temperature is raised.

The Langmuir isotherm was developed by Irving Langmuir in 1916 to describe the dependence of the surface coverge of an adsorbed gas on the pressure of the gas above the surface at a fixed temperature. There are many other types of isotherm (Temkin, Freundlich...) which differ in one or more of the assumptions made in deriving the expression for the surface coverage; in particular, on how they treat the surface coverage dependence of the enthalpy of adsorption. Whilst the Langmuir isotherm is one of the simplest, it still provides a useful insight into the pressure dependence of the extent of surface adsorption.

Important Note - Surface Coverage and the Langmuir Isotherm

When considering adsorption isotherms it is conventional to adopt a definition of surface coverage (θ) which defines the maximum *(saturation)* surface coverage of a particular adsorbate on a given surface always to be unity, i.e. $\theta_{max} = 1$.

This way of defining the surface coverage differs from that usually adopted in surface science where the more common practice is to equate θ with the ratio of adsorbate species to surface substrate atoms (which leads to saturation coverages which are almost invariably less than unity).

Langmuir Isotherm - derivation from equilibrium considerations

We may derive the Langmuir isotherm by treating the adsorption process as we would any other equilibrium process - except in this case the equilibrium is between the gas phase molecules (M), together with vacant surface sites, and the species adsorbed on the surface. Thus, for a non-dissociative (molecular) adsorption process we consider the adsorption to be represented by the following chemical equation:

$$S - * + M_{(g)} = S - M$$

where: S-*, represents a vacant surface site.

Note - in writing this equation we are making an inherent assumption that there are a fixed number of localized surface sites present on the surface. This is the first major assumption of the Langmuir isotherm.

We may now define an equilibrium constant (K) in terms of the concentrations of "reaccents" and "products"

$$K = \frac{[S - M]}{[S - *][M]}$$

We may also note that:

- [S - M] is proportional to the surface coverage of adsorbed molecules, i.e. proportional to θ
- [S - *] is proportional to the number of vacant sites, i.e. proportional to $(1-\theta)$
- [M] is proportional to the pressure of gas, P

Hence, it is also possible to define another equilibrium constant, b, as given below:

$$b = \frac{\theta}{(1 - \theta)P}$$

Rearrangement then gives the following expression for the surface coverage

$$\theta = \frac{bP}{1+bP}$$

which is the usually form of expressing the Langmuir Isotherm.

As with all chemical reactions, the equilibrium constant, b, is both temperature-dependent and related to the Gibbs free energy and hence to the enthalpy change for the process.

Note- b is only a constant (independent of θ) if the emthalpy of adsorption is independent of coverage - this is the second major assumption of the Langmuir Isotherm.

Langmuir Isotherm - Derivation from kinetic considerations

The equilibrium that may exist between gas adsorbed on a surface and molecules in the gas phase is a dynamic state, i.e. the equilibrium represents a state in which the rate of adsorption of molecules onto the surface is exactly counterbalanced by the rate of desorption of molecules back into the gas phase. It should therefore be possible to derive an isotherm for the adsorption process simply by considering and equating the rates for these two processes.

Expressions for the rate of adsorption and rate of desorption have been derived in previous chapter.

$$R_{ads} = \frac{f(n).\ P}{\sqrt{2\pi mkT}} \exp\left(-E_a^{ads}/RT\right)$$

$$R_{des} = v.f'(n).\ \exp\left(E_a^{des}/RT\right)$$

Equating these two rates yields an equation of the form:

$$\frac{P.f(n)}{f'(n)} = C(T) \qquad \ldots (9)$$

where the terms f(n) & f'(n) contain the pre-exponential surface coverage dependence of the rates of adsorption and desorption respectively and all other factors have been taken over to the r.h.s. to give a temperature-dependent "constant" characteristic of this particular adsorption process, C(T).

We now need to make certain simplifying assumptions...the first is one of the key assumptions of the Langmuir isotherm.

Adsorption takes place only at specific localized sites on the surface and the saturation coverage corresponds to complete occupancy of these sites.

Let us initially further restrict our consideration to a simple case of **reversible molecular adsorption.** i.e.

$$S\text{-}^* = M_{(g)} = S - M$$

where S-* represents a vacant surface site and S-M the adsorption complex.

Under these circumstances it is reasonable to some coverage dependencies for rates of the two processes of the form:

Adsorption, $f(n) = c\ (1\text{-}\theta)$ i.e. proportional to the fraction of sites that are unoccupied.

Desorption, $f'(n) = c'\theta$ i.e. proportional to the fraction of sites which are occupied by adsorbed molecules.

where θ is the fraction of sites occupied at equilibrium.

Note: these coverage dependencies are exactly what would be predicted by noting that the forward and reverse processes are elementary reaction steps, in which case it follows from standard chemical kinetic theory that

1. the forward adsorption process will exhibit kinetics having a first order dependence on the concentration of vacant surface sites.

2. the reverse desorption process will exhibit kinetics having a first order dependence on the concentration of adsorbed molecules.

Substitution into equation [9] then yields

$$\frac{P.(1-\theta)}{\theta} = B(T)$$

where $B(T) = (c'/c).\ C(T)$. After rearrangement this gives the Langmuir Isotherm expression for the surface coverage.

$$\theta = \frac{bP}{1 + bP}$$

where b (= 1/B(T) is a function of temperature and contains an exponential term of the form

$$b = \ldots.. \exp\left[\ (E_a^{des} - E_a^{ads})\ /RT\right] = \ldots.. \exp\left[\ -\Delta H_{ads}\ /RT\right]$$

Consequently, *b* can only be regarded as a constant with respect to coverage if the enthalpy of adsorption is itself independent of coverage - this is the second major assumption of the Langmuir Isotherm.

Variation of Surface Coverage with Temperature and Pressure

Application of the assumptions of the Langmuir Isotherm leads to readily derivable expressions for the pressure dependence of the surface coverage - in the case of a simple, reversible molecular adsorption process the expression is

$$\theta = \frac{bP}{1 + bP} \qquad \text{where } b = b(T)$$

This is illustrated in the graph below which shows the characteristic Langmuir variation of coverage with pressure for molecular adsorption.

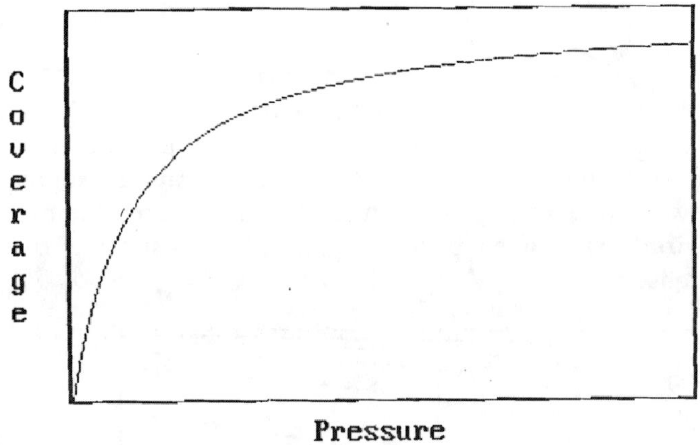

Note that:

1. $\theta \rightarrow bP$ at low pressures
2. $\theta \rightarrow 1$ at high pressures

At a given pressure the extent of adsorption is determined by the value of *b*: this in turn is dependent upon both the temperature (T) and the enthalpy (heat) of adsorption. Remember that the magnitude of the adsorption enthalpy (a negative quantity itself) reflects the strength of binding of the adsorbate to the substrate.

The value of b is increased by

1. a reduction in the system temperature
2. an increase in the strength of adsorption

Therefore the set of curves shown below illustrates the effect of either *(i)* increasing the magnitude of the adsorption enthalpy at a fixed temperature, or *(ii)* decreasing the temperature for a given adsorption system.

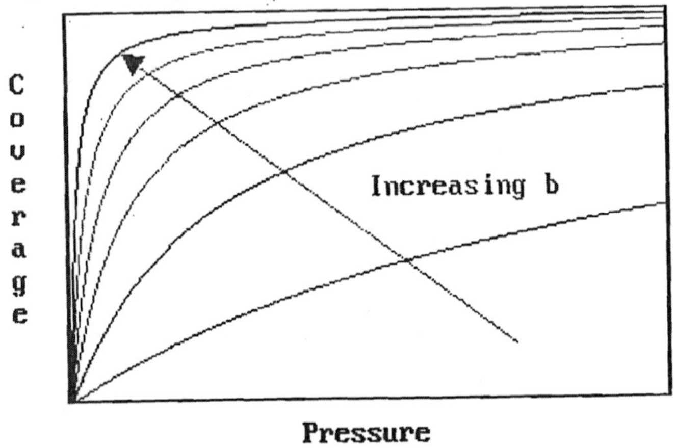

A given equilibrium surface coverage may be attainable at various combinations of pressure and temperature as highlighted below . . . note that as the temperature is lowered the pressure required to achieve a particular equilibrium surface coverage decreases.

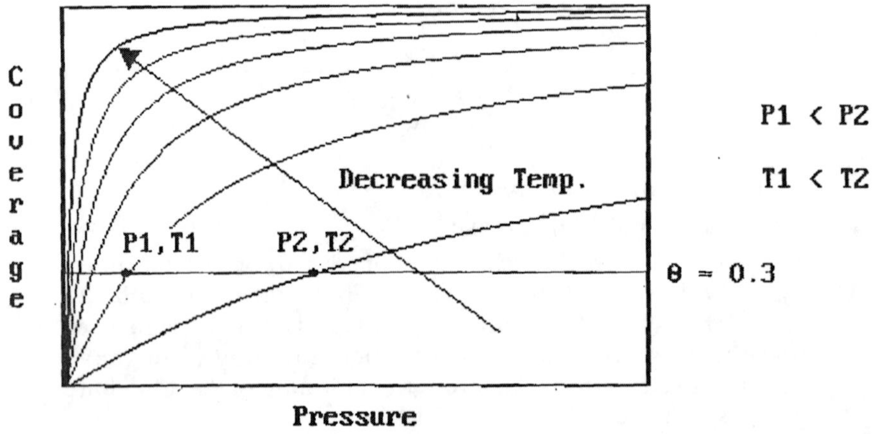

This is often used as justification for one of the main ideologies of surface chemistry; specifically, that it is possible to study technologically-important (high pressure/high temperature) surface processes within the low pressure environment of typical surface analysis systems by working at low temperatures. It must be recognised however that, at such low temperatures, kinetic restrictions that are not present at higher temperatures may become important.

If you wish to see how the various factors relating to the adsorption and desorption of molecules influence the surface coverage then try out the *Interactive Demonstration of the Langmuir Isotherm*.

Determination of Enthalpies of Adsorption

It has been shown in this how the value of b is dependent upon the enthalpy of adsorption. It has also just been demonstrated how the value of b influences the pressure/temperature (P-T) dependence of the surface coverage.

The implication of this is that it must be possible to determine the enthalpy of adsorption for a particular adsorbate/substrate system by studying the P-T dependence of the surface coverage.

Various methods based upon this idea have been developed for the determination of adsorption enthalpies - one method is outlined below:

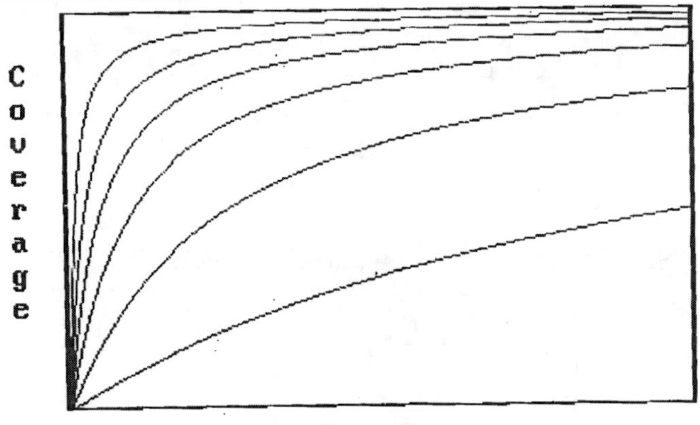

Step 1: Involves determination of a number of adsorption isotherms (where a single isotherm is a coverage/ pressure curve at a fixed temperature).

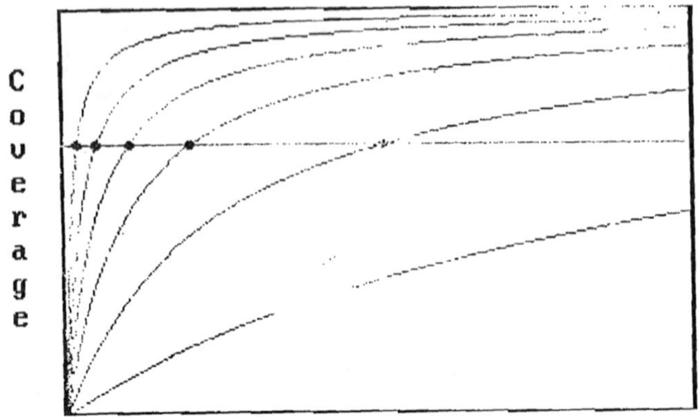

Step 2: It is then possible to read off a number of pairs of values of pressure and temperature which yield the same surface coverage.

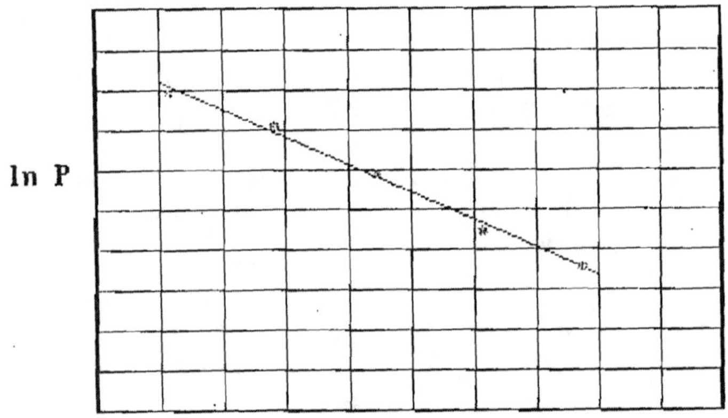

Step 3: The Clausius-Clapeyron equation

$$\left(\frac{\partial \ln P}{\partial 1/T}\right)_{const\ \theta} = \frac{\Delta H_{ads}}{R}$$

may then be applied to this set of (P-T) data and a plot of ((ln P) v's (1/T) should give a straight line, the slope of which yields the adsorption enthalpy.

Note: • the value obtained for the adsorption enthalpy is that pertaining at the surface coverage for which the P-T data was obtained, but steps 2 & 3 may be repeated for different surface coverages enabling the adsorption enthalpy to be determined over the whole range of coverages.

* this method is applicable only when the adsorption process is thermodynamically reversible.

Applications: Kinetics of Catalytic Reactions

It is possible to predict how the kinetics of certain heterogeneously-catalysed reactions might vary with the partial pressures of the reactant gases above the catalyst surface by using the Langmuir isotherm expression for equilibrium surface coverages.

Example 1: Unimolecular Decomposition

Consider the surface decomposition of a molecule A, i.e. the process

$$A_{(g)} \leftrightarrow A_{(ads)} \rightarrow \text{Products}$$

Let us assume that:

1. The decomposition reaction occurs uniformly across the surface sites at which molecule *A* may be adsorbed and is not restricted to a limited number of special sites.
2. The products are very weakly bound to the surface and, once formed, are rapidly desorbed.
3. The rate determining step (rds) is the surface decomposition step.

Under these circumstances, the molecules of *A* adsorbed on the surface are in equilibrium with those in the gas phase and we may predict the surface concentration of *A* from the Langmuir isotherm i.e.

$$\theta = b.P/(1+b.P)$$

The rate of the surface decomposition (and hence of the reaction) is given by an expression of the form

$$\text{Rate} = k\theta$$

(Note-we are assuming that the decomposition of A(ads) occurs in a simple unimolecular elementary reaction step and that the kinetics are first order with respect to the surface concentration of this adsorbed intermediate)

Substituting for the coverage, θ, gives us the required expression for the rate in terms of the pressure of gas above the surface

$$\text{Rate} = k\ b\ P/(1+bP)$$

It is useful to consider two extreme limits:

- Limit #1: $b.P \ll 1$; then

$$(1 + b.P) \sim 1 \quad \text{and} \quad \text{Rate} \sim k.b.P$$

i.e. a first order reaction (with respect to the partial pressure of *A*) with an apparent first order rate constant, $k' = k.b$.

This is the low pressure (or weak binding i.e. small b) limit: under these conditions the steady state surface coverage, θ, of the reactant molecule is very small.

- Limit #2: $b.P \gg 1$; then

$$(1 + b.P) \sim b.P \quad \text{and Rate} \sim k$$

i.e. a zero order reaction (with respect to the partial pressure of *A*).

This is the high pressure (or strong binding i.e. large b) limit: under these conditions the steady state surface coverage, θ, of the reactant molecule is almost unity.

In fact, the rate shows the same pressure variation as does the surface coverage (hardly surprising since it is directly proportional to θ)

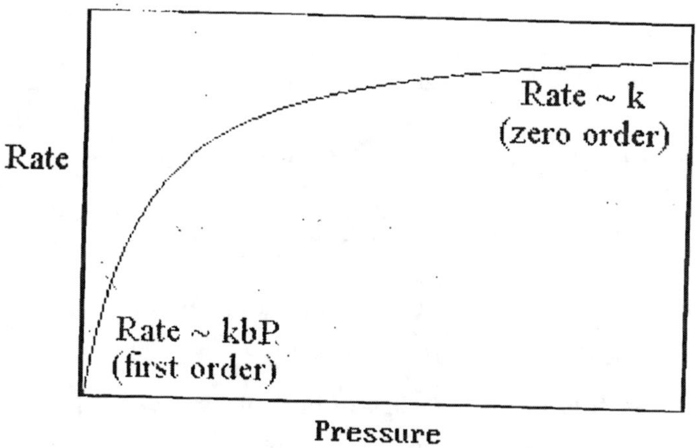

Example 2: Bimolecular Reaction

Consider a Langmuir-Hinshelwood reaction of the following type:

$$A_{(g)} = A_{(ads)}$$

$$B_{(g)} = B_{(ads)}$$

$$A_{(ads)} + B_{(ads)} \xrightarrow{\text{rids}} AB_{(ads)} \xrightarrow{\text{fast}} AB_{(g)}$$

We will further assume, as noted in the above scheme, that the surface reaction between the two adsorbed species is the rate determining step.

If the two adsorbed molecules are mobile on the surface and freely intermix then the rate of the reaction will be given by the following rate expression for the bimolecular surface combination step

$$\text{Rate} = k\theta_A\theta_B$$

For a single molecular adsorbate the surface coverage (as given by the Langmuir isotherm) is:

$$\theta = b.P/(1+b.P)$$

Where two molecules (A&B) are competing for the same adsorption sites then the relevant expressions are:

$$\theta_A = \frac{b_A P_A}{1 + b_A P_A + b_B P_B} \quad \text{and} \quad \theta_B = \frac{b_B P_B}{1 + b_A P_A + b_B P_B}$$

Substituting these into the rate expression gives:

$$\text{Rate} = k\theta_A\,\theta_B = \frac{k b_A P_A . b_B P_B}{(1+b_A P_A + b_B P_B)^2}$$

Once again, it is interesting to look at several extreme limits

* Limits #1: $b_A P_A << 1$ & $b_B P_B << 1$

In this limit θ_A & θ_B are both very low, and

Rate $\rightarrow k b_A P_A b_B P_B = k' P_A P_B$

i.e. *first order in both reactants*

* Limit #2 : $b_A P_A << 1 << b_B P_B$

In this limit $\theta_A \rightarrow 0$, $\theta_B \rightarrow 1$, and

Rate $\rightarrow k b_A P_A/(b_B P_B) = k' P_A/P_B$

i.e. *first order* in A, but *negative first order* in B.

From this example, it is readily seen that, depending upon the partial pressure and binding strength of the reactants, a given model for the reaction scheme can give rise to a variety of apparent kinetics: this highlights the dangers inherent in the reverse process - namely trying to use kinetic data to obtain information about the reaction mechanism.

BET Isotherm for Multilayer Coverage

Physical adsorption does not necessarily to with monolayer coverage of a solid surface. Instead, the newly formed layer provides adsorption sites for gas molecules, and the process of adsorption on top of adsorbed molecules continues until several layers are built. S. Brunauer, P.H. Emmett, and E. Teller developed an isotherm equation (the BET isotherm), along the following lines that describes the multilayer adsorption, Figure 8.3 is an idealized picture of a surface and a few layers of adsorbed molecules.

The number of molecules adsorbed at all sites is given by

$$n = \sum_j j s_j \qquad \ldots (10)$$

where s_j is the number of sites, each with j molecules. The number of molecules in the first layer at saturation (which is the same as the number of sites for monolayer coverage) is given by

$$n_o = \sum_j s_j \qquad \ldots (11)$$

(n_o = 7 + 4+3+0+2=16.) The BET isotherm relates n/n_o to the pressure of the gas by starting with the Langmuir isotherm. In the Langmuir theory, θ is proportional to the number of sites with one molecule each, and $(1 - \theta)$, to the number of sites with no molecules. Hence we may write as follows:

$$\frac{S_1}{S_0} = KP \qquad \ldots (12)$$

We see from this that the Langmuir equation is a mathematical statement of equilibrium between sites of different occupancy. The BET theory extends this idea by postulating equilibria between sites with different numbers of molecules. In particular it postulates

$$\frac{s_{j+1}}{s_j} = K'P \qquad \ldots (13)$$

where j = 1, 2, 3, ... Keep in mind that the equilibria are between sites with different occupancies and not between different layers. For example, when j = 4, the equilibrium is between sites with four molecules — *tetramers* — and sites with

○ Molecule

□ Site

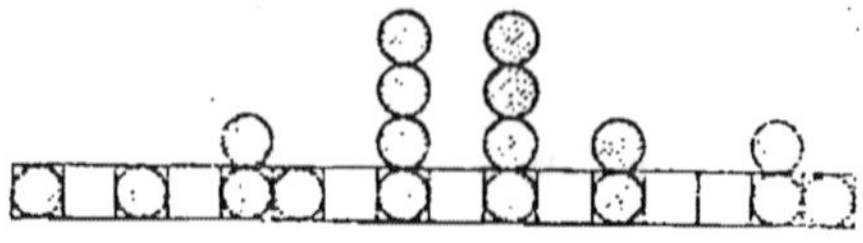

Figure 8.3: Model illustrating multilayer adsorption

five molecules —*pentamers* —and not between the molecules in the fourth and fifth layers. The BET theory assumes that the equilibrium constant for every j, other than j = 0, is the same. We have used the symbol K for the equilibrium constant between empty sites and singly occupied sites, and K′ for all subsequent equilibrium constants.

From Eq. (13) it follows that

$$s_j = K'Ps_{j-1} = (K'P)^2 s_{j-2}$$
$$= (K'P)^{j-1} s_1$$
$$= KP(K'P)^{j-1} s_o \quad \ldots (14)$$

If we define

$$K = K'c \quad \ldots (15)$$

Where c is a constant and

$$K'P = x \quad \ldots (16)$$

Eq. (14) may be written in a simpler form as

$$s_j + cx^j s_0 \quad \ldots (17)$$

The total number of molecules adsorbed [Eq. (10)] is given by

$$n = \sum_{j=0}^{\infty} js_j = cs_0 \sum_j jx^j \quad \ldots (18)$$

The sum in this expression is given by

$$\sum_{j=0}^{\infty} jx^j = \frac{x}{(1-x)^2} \quad \ldots (19)$$

Hence

$$n = \frac{cs_0 x}{(1-x)^2} \quad \ldots (20)$$

The number of sites on the surface [Eq. (11)] is given by

$$n_0 = \sum_{j=0}^{\infty} sj = s_0 + \sum_{i-1}^{\infty} si$$
$$= s_0 + cs_0 \sum x^j \qquad \ldots (21)$$

which becomes

$$n_0 = \frac{s_0 + cs_0 x}{1-x} \qquad \ldots (22)$$

when the sum is evaluated. From Eqs. (20) and (22) we get

$$\frac{n}{n_0} = \frac{cx}{(1-x)(1-x+cx)} \qquad \ldots (23)$$

A plot of the left-hand side of Eq. (23) against P/P_0 should give a straight line as shown in Fig. (8.4). The constants c and V_0 are obtained from the slope and the intercept as illustrated in the following example.

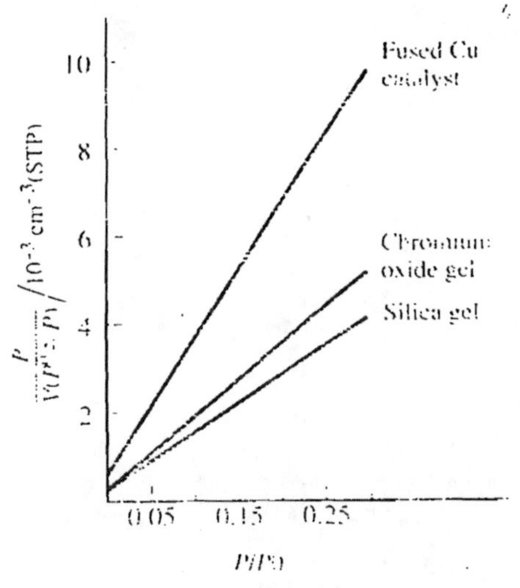

Figure 8.4: Adsorption of nitrogen at —193°C.

Example 3. From the adsorption data for nitrogen on fused copper catalyst estimate the constant *c* and the number of molecules in the first layer at saturation.

Solution: From the intercept:

$$\frac{1}{cV_o} = 5 \times 10^{-4}\ cm^{-3}$$

From the slpe:

$$\frac{c - 1}{cV_0} = 3.4 \times 10^{-2}\ cm^{-3}$$

$$c - 1 = slope/intercept = 68$$

Therefore, c =69 and (from the intercept) V_0 = 29 cm³. The number of molecules in the first layer (from the ideal gas law and V_0) = 7.79 x 10^{20}.

The V_0 values from the BET equation are used in determining the surface area of solids. The area of a nitrogen molecule is estimated to be 16 x 10^{-20} m². The surface area for fused copper catalyst is therefore given by

$$(16 \times 10^{-20}\ m^2\ molecule^{-1})(7.79 \times 10^{20}\ molecules) = 124 m^2.$$

9

Chemical Reactions on Surfaces

Introduction

An important concept in connection with surface reactions is the *molecularity*, which is the number of reactant molecules that come together during the course of reaction; we do not count the surface sites. The molecularity of a surface reaction is deduced from the kinetics on the basis of the experimental results and of theoretical considerations. Reactions involving a single reacting substance are usually, but not invariably, unimolecular. The mechanism of the surface-catalysed ammonia decomposition is, for example, usually unimolecular. On the other hand, the kinetics of the decomposition of acetaldehyde on various surfaces can only be interpreted on the hypothesis that two accetaldehyde molecules, adsorbed on neighboring surface sites, undergo a bimolecular reaction. Reactions involving two reacting substances are usually bi-molecular. When reactant molecules are dissociated on the surface, the reaction may involve interaction between an atom or radical and a molecule; for example, the exchange reaction between ammonia and deuterium on iron is a bimolecular interaction between a deuterium atom and an ammonia molecule.

Unimolecular Reactions

Surface reactions involving one molcule may be treated in terms of the Langmuir adsorption isotherm. In the simplest case the rate of reaction is proportional to θ and is thus

$$v = k\theta = \frac{kK[A]}{1 + K\,[A]} \qquad (1)$$

This equation is of the same form as the Michaelis-Menten equation, but we should note that it has been derived on the assumption of a rapid adsorption equilibrium followed by slow chemical reaction.

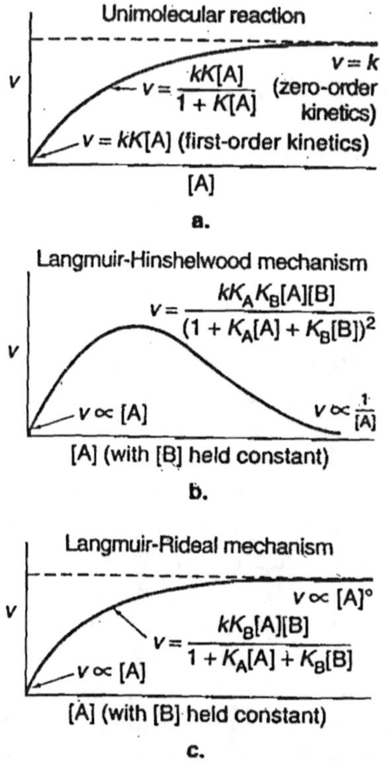

Fig. 9.1: The variation of rate with concentration for various types of surface reactions. *(a)* Simple unimolecular processes. *(b)* A bimolecular reaction occurring by a Langmuir-Hinshelwood mechanism. *(c)* A bimolecular reaction occurring by a Langmuir-Rideal mechanism.

The dependence of v on [A], shown in Figure 9.1(a), is exactly the same as that given by the Langmuir isotherm. At sufficiently high concentrations the rate is independent of the concentration, which means that the kinetics are zero order. At low concentrations, when K[A] << 1, the kinetics are first order. A good example of this type of behaviour is the decomposition of ammonia into nitrogen and hydrogen on a tungsten surface.

Sometimes a subtance other than the reactant is adsorbed on the surface, with the result that the effective surface area, and therefore the rate, are reduced. Suppose that a substance A is undergoing a unimolecular reaction on a surface and that a nonreacting substance I, known as an *inhibitor* or a *poison,* is also adsorbed. If the fraction of the surface covered by A is θ and that covered by I is θ_i, we have

$$\theta = \frac{K[A]}{1 + K[A] + K_i[I]} \tag{2}$$

where K and K_i are the adsorption constants for A and I. The rate of reaction, equal to $k\theta$, is thus

$$v = \frac{kK[A]}{1 + K[A] + K_i[I]} \tag{3}$$

In the absence of inhibitor this equation reduces to Eq. (1).

A case of special interest is when the surface is only sparsely covered by the reactant but is fairly fully covered by the inhibitor. In other words

$$K_i[I] >> 1 + K[A] \tag{4}$$

and the rate is then

$$v = \frac{kK[A]}{K_i[I]} \tag{5}$$

We can easily understand this relationship if we remember that when a surface is sparsely covered (as it now is by A), the coverage is proportional to [A]; whereas if it is almost fully covered (as it now is by I), the fraction bare is inversely proportional to [I]. A good example of Eq. (5) is provided by the decomposition of ammonia on platinum the rate law for which is

$$v = \frac{k[NH_3]}{[H_2]} \tag{6}$$

Since hydrogen is a product of reaction, there is progressive inhibition as reaction proceeds. There is no appreciable inhibition by the other reaction product, nitrogen.

Bimolecular Reactions

When a solid surface catalyses a bimolecular process, there are two possible mechanisms. One of these, first suggested by Langmuir and further developed by Hinshelwood, involves reaction between two molecules adsorbed on the surface. Such a mechanism may be formulated as follows.

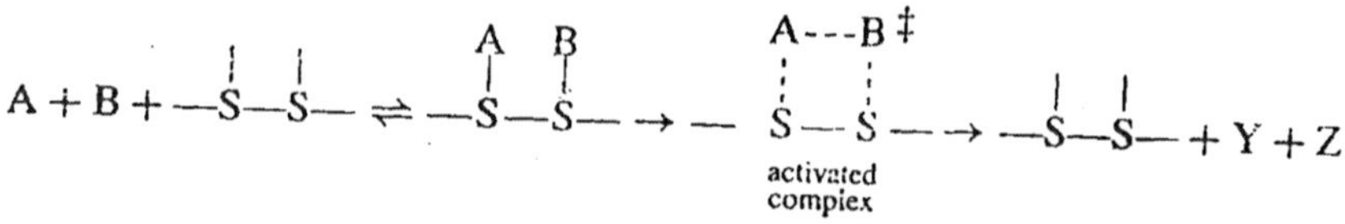

In the first step the two molecules A and B become adsorbed on neighbouring sites on the surface. Reaction then takes place, by way of an activated complex, to give the reaction products.

Langmuir-Hinshelwood Mechanism

When this Langmuir-Hinshelwood mechanism applies, the rate is proportional to the probability that A and B are adsorbed on neighbouring sites, and this is proportional to the product of fractions of the surface, θ_A and θ_B, covered by A and B. The rate of reaction is

$$v = k\theta_A\theta_B \tag{7}$$

$$= \frac{kK_AK_B\,[A]\,[B]}{(1 + K_A\,[A] + K_B[B])^2} \tag{8}$$

If [B] is held constant and [A] varied. The rate first rises, passes through a maximum, and then decreases toward zero. The physical explanation of the falling off of the rate at high concentrations is that one reactant displaces the other as its concentration is increased. The maximum rate corresponds to the presence of the maximum number of neighboring A-B pairs on the surface.

At sufficiently low concentrations of A and B, Eq. (8) predicts second-order kinetics, and this has been observed in a number of systems. Rate maxima have also often been observed. A case of special interest is when one reactant (e.g., A) is weakly adsorbed and the other strongly adsorbed. This means that $K_B[B] >> 1 + K_A[A]$, and it follows from Eq. (8) that

$$\nu = \frac{kK_A[A]}{K_B[B]} \tag{9}$$

The order is thus 1 for A and - 1 for B, an example is the reation between carbon monoxide and oxygen on quartz, wher the rate is proportional to the pressure of oxygen and inversely proportional to the pressure of carbon monoxide.

An alternative mechanism for a bimolecular surface process is for the reaction to occur between a molecule that is not adsorbed (e.g., A) and an adsorbed molecule (B). Such a mechanism was also considered by Langmuir and was further developed by the British physical chemist Sir Eric K. Rideal (1890-1974). This mechanism may be represented as

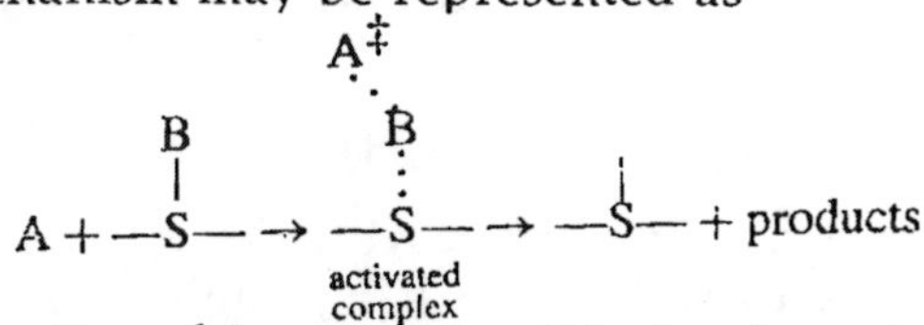

The adsorption of A may occur; it is simply postulated, in this mecahnism, that an adsorbed A does not react.

Langmuir-Rideal Mechanism

The rate for this *Langmuir-Rideal Mechanism* is proportional to the concentration of A and to the fraction of the surface that is covered by B:

$$\nu = k[A]\theta_B \tag{10}$$

$$= \frac{kK_B[A]\,[B]}{1 + K_A[A\} + K_B[B]} \tag{11}$$

The variation of v with [A] is the same type of curve is found if v is plotted against [B]. There is now no maximum in the rate, and this provides a possible way of distinguishing between this mechanism and the Langmuir-Hinshelwood Mechanism.

Not many ordinary chemical reactions occur by a Langmuir-Rideal Mechanism. There is evidence, however, that radical combinations on surfaces sometimes occur in this way. The combination of hydrogen atoms, for example, is sometimes a first-order reaction, and it appears to occur by the mechanism

$$\mathrm{H} + -\overset{\overset{\displaystyle \mathrm{H}}{|}}{\mathrm{S}}- \rightarrow -\overset{|}{\mathrm{S}}- + \mathrm{H_2}$$

The fraction of the surface covered by hydrogen atoms is

$$\theta = \frac{K[H]}{1 + K[H]} \tag{12}$$

and the rate of combination is thus

$$v = k[H]\ \theta = \frac{kK[H]^2}{1 + K[H]} \tag{13}$$

At lower temperatures the surface may be fully covered, so that $K[H] >> 1$ and

$$v = k[H] \tag{14}$$

and the kinetics are first order. At higher temperatures the coverage decreases and if $1 >> K[H]$, the rate is

$$v = kK[H]^2 \tag{15}$$

An increase in order from 1 to 2 has in fact been observed as the temperature is raised.

In 1912, and later, Langmuir made important investigations on the production of hydrogen atoms at hot tungsten surfaces, such as used in tungsten-filament incandescent lamps. The rate is proportional to the square root of the hydrogen pressure, and the mechanism is believed to be the reverse of that just given, namely

$$\mathrm{H_2} + -\overset{|}{\mathrm{S}}- \rightarrow -\overset{\overset{\displaystyle \mathrm{H}}{|}}{\mathrm{S}}- + \mathrm{H}$$

10

Residence Times

Introduction

Since the equilibrium between gaseous and adsorbed phases must represent a balance between adsorption and desorption rates, then there must be a finite residence time for an adsorbed molecule on the surface. In 1924, Frenkel proposed that this could be expressed as

$$\tau = \tau_o \exp (q_{st}/RT) \tag{1}$$

where τ_o is the oscillation time for vibration of the adsorbed molecule perpendicular to the surface. However, the problem is perhaps best treated by statistical thermodynamic means.

At adsorption equilibrium for a single gas, the chemical potentials of the adsorbed and gaseous phases must be equal. If the number of molecules in the adsorbed and gaseous states are N_a and N_g respectively, and $_af$ and $_gf$ are the corresponding partition functions, then

$$\frac{_gf}{N_g} = \frac{_af}{N_a} \exp (q_{st}/RT) \tag{2}$$

For each partition function there are possible translational, vibrational and rotational contributions, thus equation (2) may be written

$$\frac{_gf_{tr}\ _gf_{vib}\ _gf_{rot}}{N_g} = \frac{_af_{tr}\ _af_{vib}\ _af_{rot}}{N_a} \exp (q_{st}/RT) \tag{3}$$

The translational partition functions for ideal three-dimensional and two dimensional gases are given by

$$ {}_{g}f_{tr} = \left[\frac{2\pi mkT}{h^2}\right]^{3/2} \frac{N_g kT}{P} \tag{4} $$

and

$$ {}_{a}f_{tr} = \frac{2\pi m\ kT}{h^2}\ S \tag{5} $$

where S is the surface area. If, however, translation is hindered in the adsorbed state then

$$ {}_{a}f_{tr} = \frac{2\pi m\ kT}{h^2}\ S \times \frac{{}_{a}f_{tr}\ \text{(hindered)}}{{}_{a}f_{tr}\ \text{(ideal 2d)}} \tag{6} $$

If adsorption does not alter the internal vibrations of the molecule, but converts one degree of translational freedom into a vibrational mode perpendicular to the surface, then

$$ {}_{a}f_{vib} = {}_{g}f_{vib} \cdot f_z \tag{7} $$

where f_z is the vibrational partition function for the additional vibration. Combining equations (3) (7) gives

$$ \frac{N_a}{S} = \frac{P}{(2\pi m\ kT)^{1/2}} \frac{h}{kT} . f_z . \frac{{}_{a}f_{tr}\ \text{(hindered)}}{{}_{a}f_{tr}\ \text{(ideal 2d)}} . \frac{{}_{a}f_{rot}}{{}_{g}f_{rot}} \exp(q_{st}/RT) \tag{8} $$

Comparing this equation with equation (1) indicates that

$$ \tau_o = \frac{h}{kT} f_z \frac{{}_{a}f_{tr}\ \text{(hindered)}}{{}_{a}f_{tr}\ \text{(ideal 2d)}} . \frac{{}_{a}f_{rot}}{{}_{g}f_{rot}} \tag{9} $$

Thus for the case where rotation of the molecule is unrestricted on adsorption (for example, a monatomic gas) and the film is truly mobile then

$$ \tau_o = \frac{h}{kT} f_z \tag{10} $$

It is possible therefore to calculate a value for τ_o if a value for f_z can be determined. If the vibration perpendicular to the surface is excited, the frequency is low, and the partition function will be given by $f_z = kT/hv_z$, then $\tau o = 1/v_z$, which would be consistent with Frenkel's original ideas. However, if the vibrational mode is unexcited, then $f_z = 1$ and

$$ \tau_o = h/kT \tag{11} $$

If translation on the surface is hindered then equation (9) would indicate a lower value of τ_o, and in the extreme case of localised adsorption the derivation would have to be modified to account for entropy changes associated with distributing N_a molecules over N_s adsorption sites. In this case, if the perpendicular vibrational mode is again unexcited, then

$$\tau_o = \frac{h}{kT} \cdot \frac{h^2 N_s/S}{2\pi m\, kT} \cdot \frac{{}_a f_{rot}}{{}_g f_{rot}} \tag{12}$$

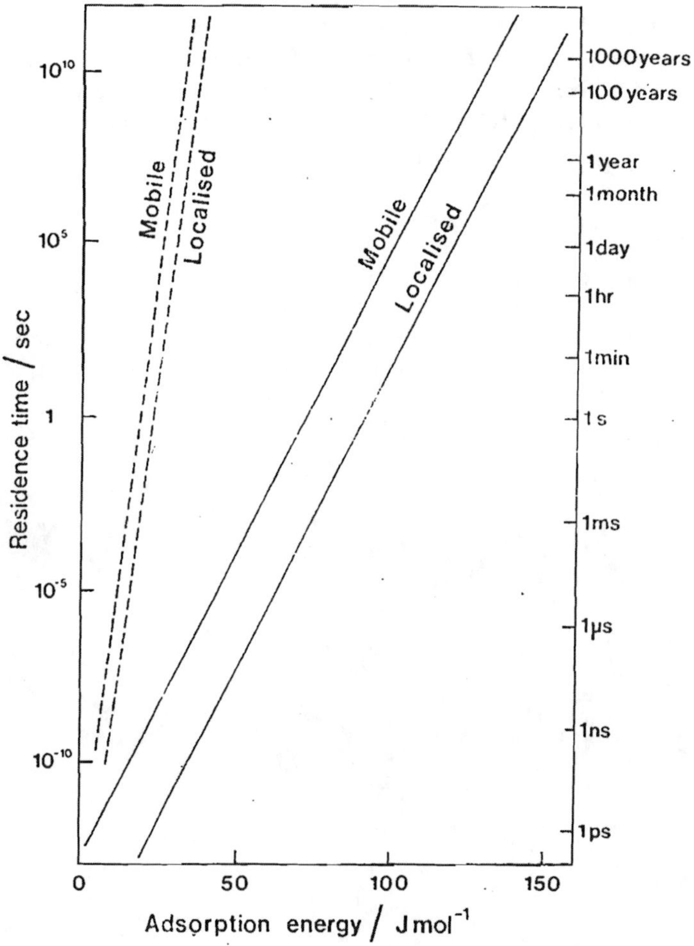

Fig.10.1: Approximate residence times of molecules in mobile and localised adsorbed films as a function of adsorption energy, at

De Boer has suggested that for the localised adsorption of water on charcoal at 300 K, the value of τ_0 is ~ 10^{-16}s.

To assess the implications of this approach on residence time, it is pertinent to calculte τ at various temperatures, but to do this we need some assumptions about the change in τ_0 for a localised film with temperature. If the ratio of rotational partition functions is unaltered, then equation (10) indicates the $\tau_0 \, \alpha T^{-2}$. Thus at 77.5 K the value of τ_0 for a localised film is ~ 1.5×10^{-15}s.

Fig. (10.1) shows approximate values of τ calculated using the value of τ_0 from equation (9) for mobile films (1.6×10^{-13} s and 6.2×10^{-13} s at 298.15 and 77.5 K respectively, and the values quoted in the preceding paragraph for localised films.

11

Surface Sensitivity and Surface Specificity

1. General Sensitivity Problems

The problems of sensitivity and detection limits are common to all forms of spectroscopy; some techniques are simply better than others in this respect ! In its simplest form, the question of sensitivity boils down to whether it is possible to detect the desired signal above the noise level.

In virtually all surface studies (especially those on single crystal substrates) sensitivity is a major problem. Consider the case of a sample with a surface of size 1 cm^2 - this will have ca. 10^{15} atoms in the surface layer. In order to detect the presence of impurity atoms present at the 1 % level, a technique must be sensitive to ca. 10^{13} atoms.

Contrast this with a spectroscopic technique used to analyze a 1 cm^3 bulk liquid sample i.e. a sample of ca. 10^{22} molecules. The detection of 10^{13} molecules in this sample would require 1 ppb (one *part-per-billion*) sensitivity - very few techniques can provide anything like this level of sensitivity !

This is one reason why common spectroscopic techniques such as NMR (detection limit ca. 10^{19} molecules) cannot be used for surface studies, except on samples with very high surface areas.

2. Surface Sensitivity and Specificity

Assuming that a technique of sufficient sensitivity can

be found, another major problem that needs to be addressed in surface spectroscopy is distinguishing between singnals from the surface and the bulk of the sample. In principle, there are two ways around this problem.

1. To ensure that the surface signal is distinguishable (shifted) from the comparable bulk signal, and that the detection system has sufficient dynamic range to detect very small signals in the presence of neighbouring large signals.

2. To ensure that the bulk signal is small compared to the surface signal i.e. that the vast majority of detected signal comes from the surface region of the sample.

It is the latter approach which is used by the majority of surface spectroscopic techniques - such techniques can then be said to be *surface sensitive.* To illustrate this, we shall look at one way in which surface specificity can be achieved - this makes use of the special properties of

..... Low Energy Electrons

— It is an approach employed in common surface spectroscopic techniques such as

Auger Electron Spectroscopy (AES)

X-ray Photoelectron Spectroscopy (XPS)

We shall illustrate this approach by considering the latter technique (XPS) but you should remember that the ideas that we develop are more widely applicable.

XPS - A surface sensitive technique?

In answering this important question, we will actually address the topics listed below - you are strongly advised to work through them in the order in which they are given.

Contents: What do we really mean by surface sensitivity/ specificity?

How can we demonstrate that XPS is surface sensitive?

Why is the XPS technique surface sensitive?

How can we use knowledge of the IMFP to calculate the thickness of surface films?

How can we change the degree of surface sensitivity?

What do we really mean by surface sensitivity/ specificity?

Most analytical techniques used in chemistry are "bulk" techniques in the sense that they measure all the atoms within a typical sample (be it a solid, liquid, solution or gas phase sample) - so if, for example, we were to analyse a sample consisting of a 10 nm thick layer of one material (A) deposited on a 1 mm thick substrate of another material (B)....

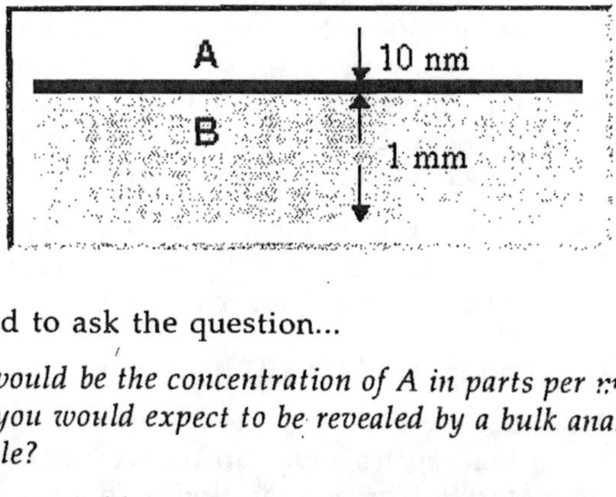

. . . and to ask the question...

What would be the concentration of A in parts per million (ppm) which you would expect to be revealed by a bulk analysis of such a sample?

. . . . we would get the same result (10 ppm) as if we were looking at a sample in which component A was uniformly distributed throughout the bulk of B such as to give the same overall concentration in the sample as a whole.

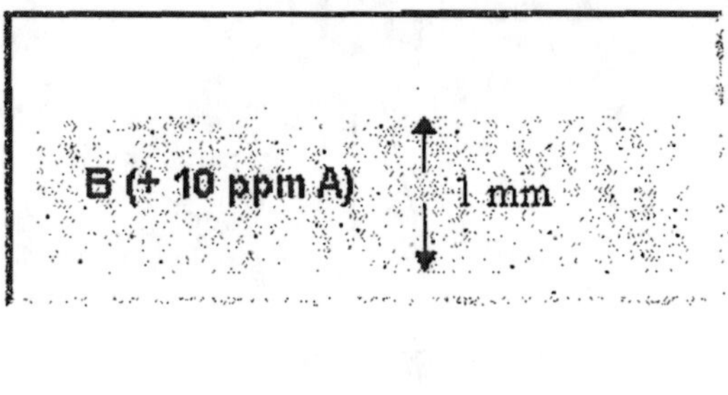

By contrast, a *surface sensitive* technique is more sensitive to those atoms which are located near the surface than it is to atoms in the bulk which are well away from the surface (i.e. the main part of the signal comes from the surface region) - in the case of the first sample therefore a surface sensitive technique would produce a spectrum with enhanced signals due to component A, i.e.

$$I_A / I_B >> 10^{-5}$$

(where I_A is the signal due to component A etc.)

A truly *surface specific* technique should **only** give signals due to atoms in the surface region - but that, of course, rather depends how the surface region is defined!

Electron spectroscopic techniques such as XPS are not completely surface specific (although you will occasionally find this expression being used) in that whilst most of the signal comes from within a few atomic layers of the surface, a small part of the signal comes from much deeper into the solid - they are best described as being surface sensitive techniques.

However, we should also note that there are a few techniques (e.g. low energy ion scattering [ISS/LEIS] which are indeed virtually specific to the topmost layer of atoms.

How can we demonstrate that XPS is surface sensitive?

Consider again the first of the two samples considered in the previous section - such samples are often readily prepared by depositing material A by evaporation onto a substrate of material B.

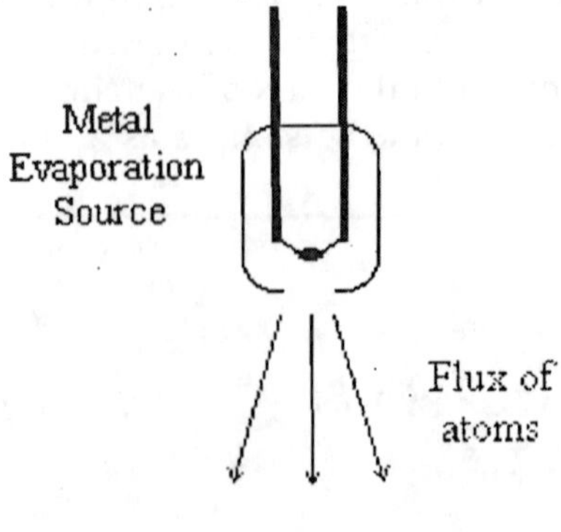

It is experimentally observed that in such a situation the XPS signals due to the substrate (material B) are rapidly attenuated, whilst those due to the condensed evaporant (material A) simultaneously increase to a limiting value.

Note: It is possible to measure deposition rates using equipment such as quartz crystal oscillators and thus obtain independent verification of the deposited film thickness. A detailed analysis of the variation of the signals due to A and B is also capable of yielding information about the mechanism of film growth.

Why is the XPS technique surface sensitive?

The soft X-rays employed in XPS penetrate a substantial distance into the sample (~μm). Thus this method of excitation imparts no surface sensitivity at the required atomic scale. The surface sensitivity must therefore arise from the emission and detection of the photoemitted electrons.

Consider, therefore, a hypothetical experiment in which electrons of a given energy, E_0, are emitted from atoms in a solid at various depths, *d*, below a flat surface.

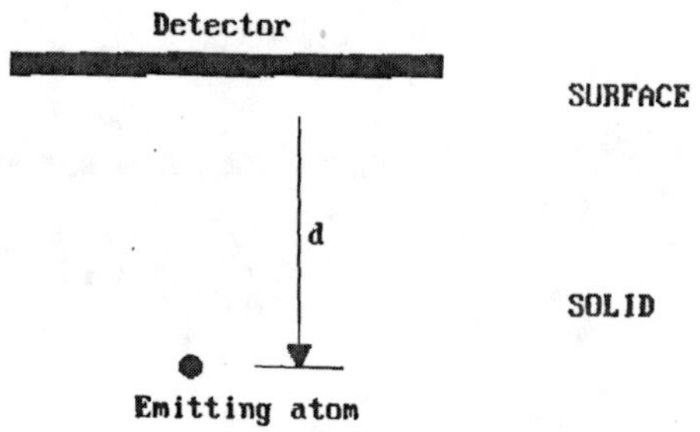

We shall assume that only those electrons which reach the surface and, at the time of leaving the solid, still have their initial energy (E_0) are detected in this experiment.

(*Note* - in an XPS experiment it is only those electrons in the peaks with the characteristics emission energies that we consider in the subsequent analysis; we do not consider electrons which have lost energy and contribute to the background of the spectrum).

There are two things which would therefore prevent an emitted electron from being detected;

1. If it were "captured" before reaching the surface, or emitted in the wrong direction and never reached the surface!

2. If it lost energy before reaching the surface, i.e. if it escaped from the solid and reached the detector, but with $E < E_o$

The process by which an electron can lose energy as it travels through the solid is known as "inelastic scattering". Each inelastic scattering event leads to : *(i)* a reduction in the electron energy *(ii)* a change in the direction of travel.

In the diagram below arrows are used to represent possible trajectories of emitted electrons from a particular source atom. The orange arrows represent electrons travelling with their initially energy, E_o, whilst the blue arrows represent electron trajectories subsequent to an inelastic scattering event. Only the orange traces which extend out of the solid therefore correspond to electrons which would be detected in our hypothetical experiment.

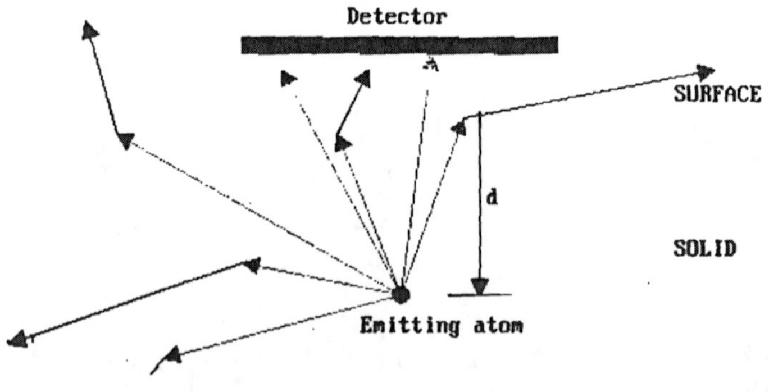

In this case, therefore, only a small fraction of the emitted electrons would have been detected.

Clearly if the source atom had been closer to the surface, then a greater fraction of the emitted electrons would have been detected since:

1. The detector would subtend a greater (solid) angle at the emitting atom (this is actually not a significant factor in real experiments).
2. For an electron emitted towards the surface, there would be less chance of inelastic scattering before it escaped from the solid.

So how does the probability of detection depend upon the distance of the emitting atom below the surface?

Let us first simplify the problem by only considering those electrons emitted directly towards (normal to) the surface - prictorially, we can represent this by a smaller detector located directly above the emitting atoms.

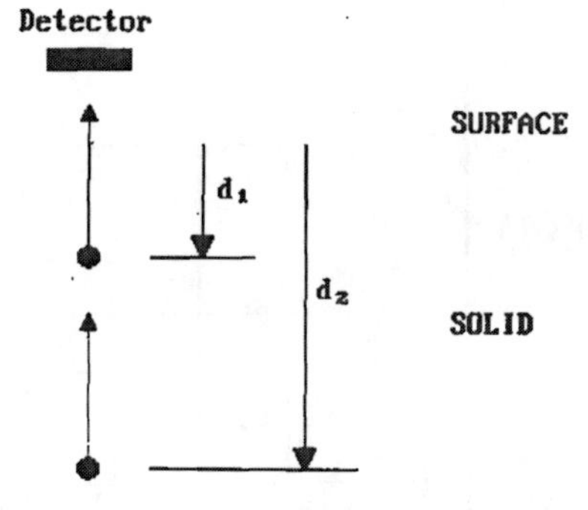

The probability of escape from a given depth, P(d), is determined by the likelihood of the electron not being inelastically scattered during its journey to the surface.

Clearly, $\mathbf{P(d_2) < P(d_1)}$ but, in order to quantify this we need to know more about the inelastic scattering process - this brings us to the subject of the...

Inelastic Mean Free Path (IMFP) of electrons.

The IMFP is a measure of the average distance travelled by an electron through a solid before it is inelastically scattered; it is dependent upon

- The initial kinetic energy of the electron.
- The nature of the solid (**but** most elements show very similar IMFP v's energy relationships).

The IMFP is actually defined by the following equation which gives the probability of the electron travelling a distance, d, through the solid without undergoing scattering

$$P(d) - \exp\left(-\ d/\ \lambda\right)$$

where λ is the IMFP for the electrons of energy E (note : $\lambda = \lambda$ (E), and that this inelastic mean free path of electrons in the solid is, of course, completely unrelated to the mean free path in the gas phase once they escape from the solid). The graph below illustrates this functional form of P(d). . .

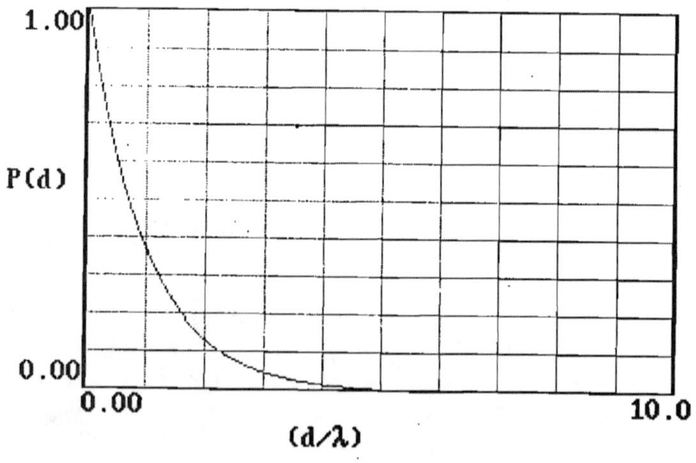

. . . and it is clear that the probability of escape decays very rapidly and is essentially zero for a distance d > 5λ. It is therefore useful to re-plot the function for a range of (d/λ) from 0-5.

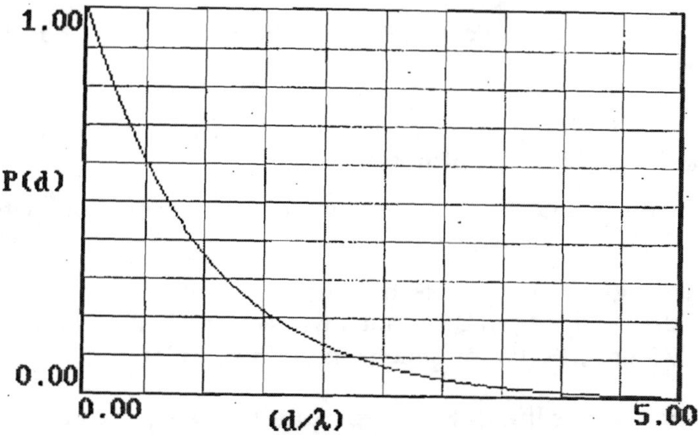

What is the probability of escape when . . . d = λ ?

What is the probability of escape when . . . d = 2λ ?

What is the probability of Escape when . . . d = 3λ ?

It is also interesting to took at this same problem from a different point of view. Loet us assume that there are many sources of electrons uniformly distributed at all distances from the surface of the solid (which is in effect the situation pertaining in all XPS experiments) and again arrange to only detect those unscattered electrons which emerge normal to the surface - then we can ask...

What is the distribution of the depths of origin of those electrons that we do detect?

This new function, let's call it P' (d), actually has the same exponential form as P(d) - i.e. the frequency of detection of electrons from different depths in the solid is directly proportional to the probability of electron escape from each depth.

So if we allow our detector to collect 100 electrons, we can ask...

How many electrons (on average) will have come from within a distance of one IMFP from the surface?

(i.e. how many originate from a distance, d, such that $0 < (d/\lambda) < 1$?).

This is a relatively simple problem in integration-try it!

In summary, we can conclude from the answer to this question that...

. . . the majority of electrons detected come from within one "imfp" of the surface.

Virtually all (>95%) of the electrons detected come from within 3 IMFPs of the surface.

Warning - beware of terms such as "*sampling depth*" and "*electron escape depth*". It is not always clear whether these refer to 1, 3 or 5 IMFPs (or something else completely).

So how big is the IMFP of typical electrons within a solid such as a metal?

The graph below shows the so-called "universal curve" for the variation of the IMFP with initial electron energy (in eV).

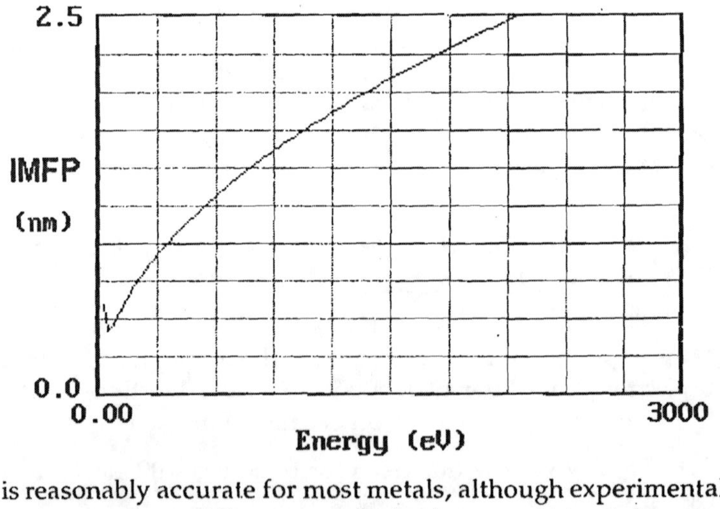

It is reasonably accurate for most metals, although experimental measurements on different metals do show a substantial scatter about the general curve.

Other classes of solids also exhibit qualitatively similar universal curves, although the absolute values of the IMFP of electrons in materials such as inorganic solids and polymers do differ quite significantly from those in metals.

This IMFP data is more usually displayed on a *log-log* plot, since this more clearly shows how the IMFP varies at low kinetic energies - the graph below shows the general curve for metals re-plotted in this new format.

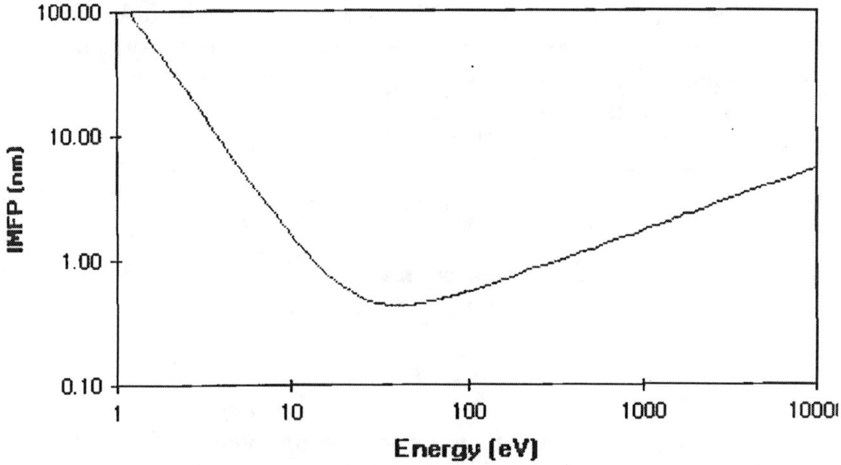

The IMFP exhibits a minimum for electrons with a kinetic energy of around 50-100 eV; at lower energies the probability of inelastic scattering decreases since the electron has insufficient energy to cause plasmon excitation (the main scattering mechanism in metals), and consequently the distance between inelastic collisions and the IMFP increases.

In summary,

The Inelastic Mean Free Path (IMFP) in metals is typically less than:

10 Angstroms (1 nm) for electron energies in the range $15 < E/eV < 350$

20 Angstroms (2 nm) for electron energies in the range $10 < E/eV < 1400$

i.e. the IMFP of low energy electrons corresponds to only a few atomic layers.

This means that any experimental technique such as XPS which involves the generation and detection of electrons of such energies will be surface sensitive.

How can we use knowledge of the IMFP to calculate the thickness of surface films?

We can now consider a situation where a substrate (or thick film) of one material, B, is covered by a thin film of a different material, A. The XPS signal from the underlying substrate will be attenuated (i.e. reduced in intensity) due to inelastic scattering of some of the photoelectrons as they traverse through the layer of material A.

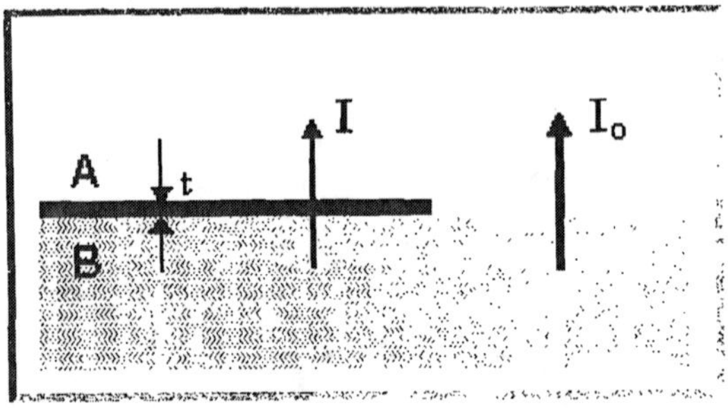

The probability of such a scattering event for any single photoelectron passing through this layer is simply given by:

$$P = \exp(-t/\lambda)$$

where: t is the thickness of the layer of material A.

If follows that the overall intensity of a XPS signal arising from B is reduced by this same factor, i.e. if the intensity of this signal in the absence of any covering layer is I_0 then the intensity I in the presence of the overlayer is given by:

$$I = I_0 \exp(-t/\lambda)$$

It also follows that it is possible to estimate the thickness of a deposited layer (using the above equation), provided the reduction in the substrate signal is known (i.e. if spectra are acquired before and after deposition of the covering film).

How can we change the degree of surface sensitivity?

For a given technique such as XPS which relies on electron scattering to provide the surface sensitivity, we can consider how it might be possible to increase the surface sensitivity when analysing a particular substrate.

One approach would be to try and ensure that the electrons being detected have kinetic energies corresponding to the minimum in the universal curve for the IMFP. The problem with this approach is that

1. it requires us to be able to vary the X-ray radiation so that we can adjust the photon energy, hv, to give us the desired kinetic energy for the photoemitted electrons; this is *not* generally possible with conventional laboratory x-ray sources (although tunable x-ray radiation is available at large synchrotron sources).

2. whilst it may be possible to adjust the kinetic energy of some particular photoelectrons to correspond to the minimum in the universal curve, this effect may not hold for all the other photoelectron peaks associated with the sample.

A much more generally useful approach is to collect photoelectrons at a more grazing emission angle.

The key point to recognize here is that it is actually the distance that the electrons have to travel through the solid (rather than the distance of the emitting atom below the surface) that is crucial in determining their probability of escape without inelastic scattering—only in the special case considered previously of normal emission are these two quantities the same.

Consider, for example, emission at some angle θ to the surface normal.

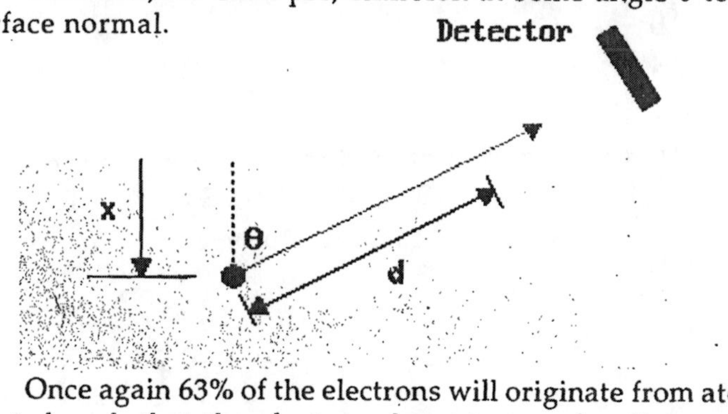

Once again 63% of the electrons will originate from atoms located such that the electrons have to travel a distance, d, which is less than 1λ through the solid (see above). Under these measurement conditions, however, all these atoms are located within a smaller distance, x, from the surface where

$$x = d \cos \theta$$

So as the emission angle (θ) is increased, $\cos \theta$ decreases, the analysed region becomes more surface localised and the surface sensitivity is increased.

One example of how this effect can be put to good use is in the study (and diagnosis) of *surface segregation*. The graph below shows simulated data for the variation of the XPS signal intensities with emission angle for

- a random alloy of 10% of element A in element B, with no segregation (blue curve).
- the same random alloy, but with surface segregation of A to give a surface monolayer of pure element A (red curve).

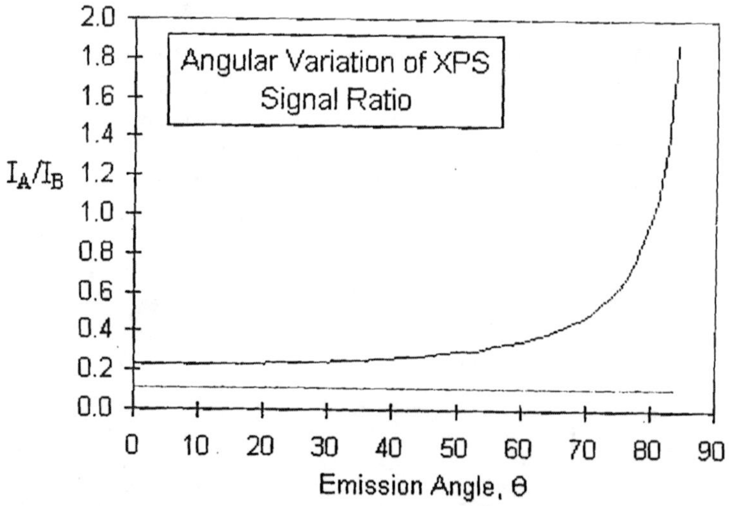

12

Auger Electron Spectroscopy

Introduction

Auger Electron Spectroscopy *(Auger spectroscopy or AES)* was developed in the late 1960's. deriving its name from the effect first observed by Pierre Auger, a French Physicist, in the mid-1920's. It is a surface specific technique utilising the emission of low energy electrons in the *Auger process* and is one of the most commonly employed surface analytical techniques for determining the composition of the surface layers of a sample.

PHYSICAL BASIS

Auger spectroscopy can be considered as involving three basic steps:

(1) Atomic ionisation (by removal of a core electron)

(2) Electron emission (the Auger process)

(3) Analysis of the emitted Auger electrons

The last stage is simply a technical problem of detecting charged particles with high sensitivity, with the additional requirement that the kinetic energies of emitted electrons must be determined.

We will therefore concern ourselves only with the first two processes - before starting, however, it is useful to briefly review the electronic structure of atoms and solids, and associated nomenclature.

Electronic Structure - Isolated Atoms

The diagram below schematically illustrates the energies of the various electron energy levels in an isolated, multi-electron atom, with the conventional chemical nomenclature for these orbitals given on the right hand side.

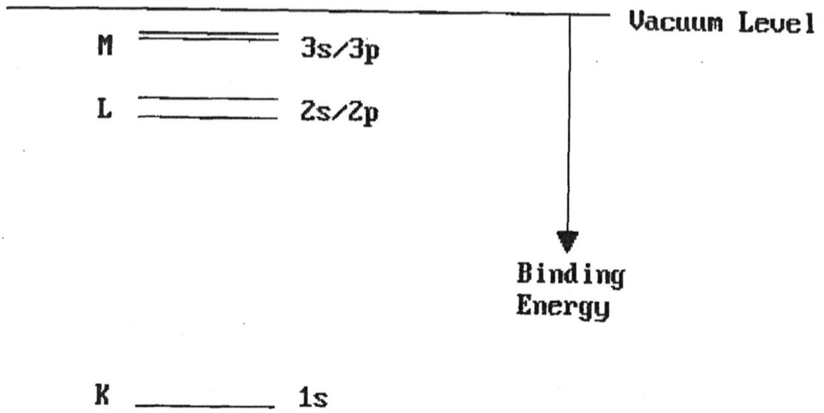

However, scientists working with x-rays tend to use the alternative nomenclature on the left and it is this nomenclature that is used in Auger spectroscopy. The designation of levels to the K, L, M, ... shells is based on their having principal quantum numbers of 1,2,3,... respectively.

It is convenient to expand the part of the energy scale close to the vacuum level in order to more clearly distinguish between the higher levels...

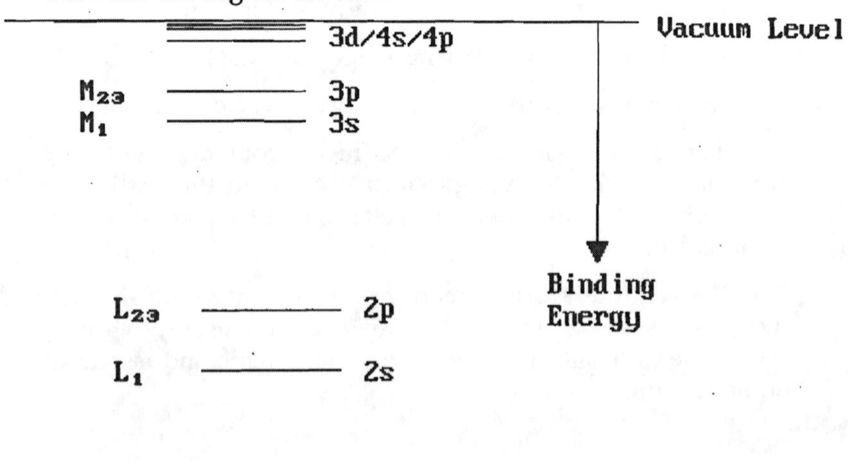

The numberical component of the KLM . . . style of nomenclature is usually written as a subscript immediately following the main shell designation.

Levels with a non-zero value of the orbital angular momentum quantum number (l >0), i.e. p,d,f,.. levels, show spin-orbit splitting. The magnitude of this splitting, however, is too small to be evident on this diagram-hence, the double subscript for these levels (i.e. $L_{2,3}$ respresents both the L_2 and L_3 levels).

Electronic Structure - Solid State

In the solid state the core levels of atoms are little perturbed and essentially remain as discrete, localised (i.e. atomic-like) levels. The valence orbitals, however, overlap significantly with those of neighbouring atoms generating bands of spatially-delocalised energy levels.

The energy level diagram for the solid is therefore closely resemblant of that of the corresponding isolated atom, except for the levels closest to the vacuum level.

The diagram below shows the electronic structure of Na metal:

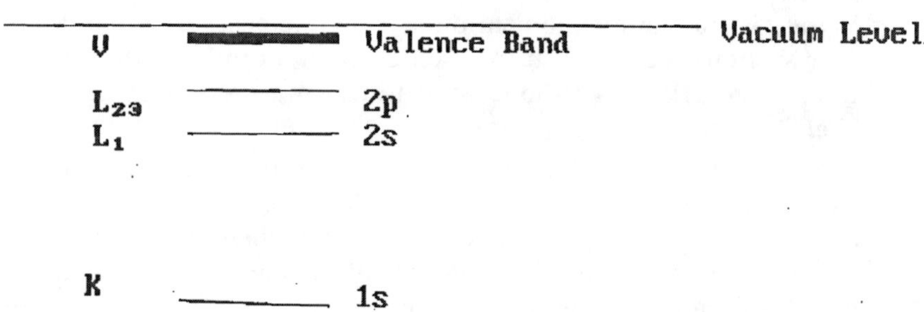

The Auguer Process & Auger Spectroscopy

Now let us return to the subject of Auger spectroscopy - in the following discussion, the Auger process is illustrated using the K, L_1 & $L_{2,3}$ levels. These could be the inner core levels of an atom in either a molecular or solid-state environment.

I. Ionisation

The Auger process is initiated by creation of a core hole - this is typically carried out by exposing the sample to a beam of high energy electrons (typically having a primary energy in the range 2 - 10 keV). Such electrons have sufficient energy to ionise all levels of the lighter elements, and higher core levels of the heavier elements.

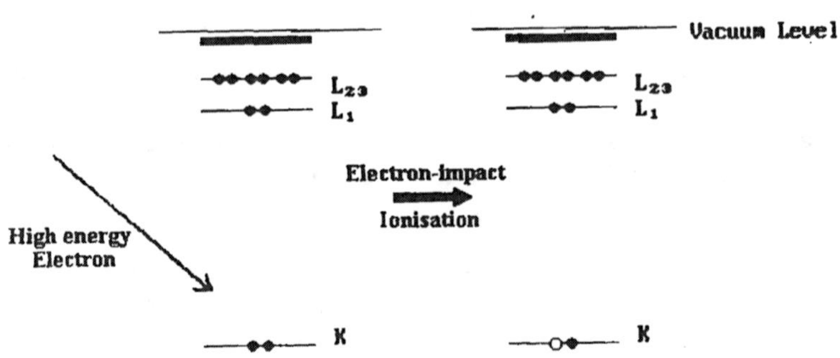

In the diagram above, ionisation is shown to occur by removal of a K-shell electron, but in practice such a crude method of ionisation will lead to ions with holes in a variety of inner shell levels.

In some studies, the initial ionisation process is instead carried out using soft x-rays (hv = 1000 - 2000 eV). In this case, the acronym XAES is sometimes used. As we shall see, however, this change in the method of ionisation has no significant effect on the final Auger spectrum.

II. Relaxation and Auger Emission

The ionised atom that remains after the removal of the core hole electron is, of course, in a highly excited state and will rapidly relax back to a lower energy state by one of two routes:

1. X-ray fluorescence, or
2. Auger emission

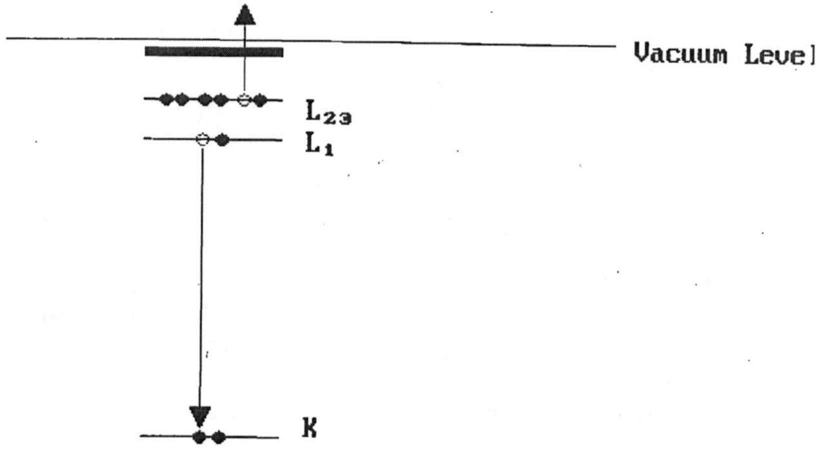

We will only consider the latter mechanism, an example of which is illustrated schematically below...

In this example, one electron falls from a higher level to fill an initial core hole in the K-shell and the energy liberated in this process is simultaneously transferred to a second electron; a fraction of this energy is required to overcome the binding energy of this second electron, the remainder is retained by this emitted *Auger electron* as kinetic energy. In the Auger process illustrated, the final state is a doubly-ionised atom with core holes in the L_1 and $L_{2,3}$ shells.

We can make a rough estimate of the KE of the Auger electron from the binding energies of the various levels involved. In this particular example,

$$KE = (E_K - E_{L1}) - E_{L23}$$

[Why is this answer not likely to be very accurate ?]

Note: the KE of the Auger electron is idependent of the mechanism of initial core hole formation.

The expression for the energy can also be re-written in the form:

$$KE = E_K - (E_{L1} + E_{L23})$$

It should be clear from this expression that the latter two energy terms could be interchanged without any effect - i.e. it is actually impossible to say which electron fills the initial core hole and which is ejected as an Auger electron; they are indistinguishable.

An Auger transition is therefore characterised primarily by:

1. the location of the initial hole
2. the location of the final two holes

although the existence of different electronic states (terms) of the final doubly-ionised atom may lead to fine structure in high resolution spectra.

When describing the transition, the initial hole location is given first, followed by the locations of the final two holes in order of decreasing binding energy.

i.e. the transition illustrated is a $KL_1L_{2,3}$ transition.

If we just consider these three electronic levels there are clearly several possible Auger transitions: specifically,

KL_1L_1 $\qquad$ $KL_1L_{2,3}$ $\qquad$ $KL_{2,3}\ L_{2,3}$

In general, since the initial ionisation is non-selective and the initial hole may therefore be in various shells, there will be many possible Auger transitions for a given element - some weak, some strong in intensity. AUGER SPECTROSCOPY is based upon the measurement of the kinetic energies of the emitted electrons. Each element in a sample beign studied will give rise to a characteristic spectrum of peaks at various kinetic energies.

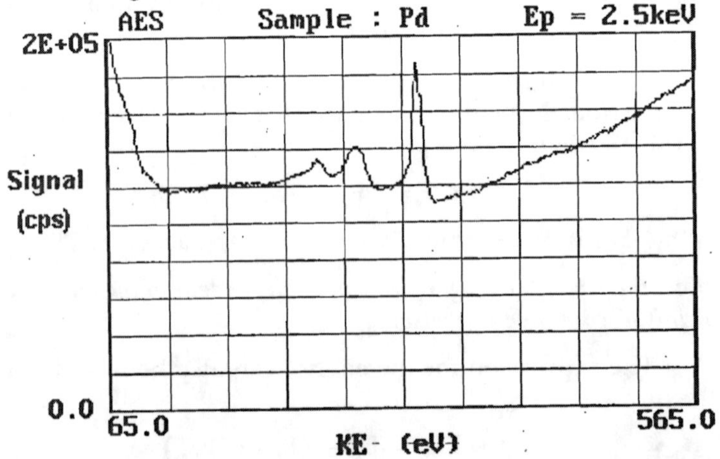

This is an Auger spectrum of Pd metal - generated using a 2.5 keV electron beam to produce the initial core vacancies and hence to stimulate the Auger emission process. The main peaks for palladium occur between 220 & 340 eV. The peaks are situated on a high background which arises from the vast number of so-called *secondary electrons* generated by a multitude of inelastic scattering processes.

Auger spectra are also often shown in a differentiated form: the reasons for this are partly historical, partly because it is possible to actually measures spectra directly in this form and by doing so get a better sensitivity for detection. The plot below shows the same spectrum in such a differentiated form.

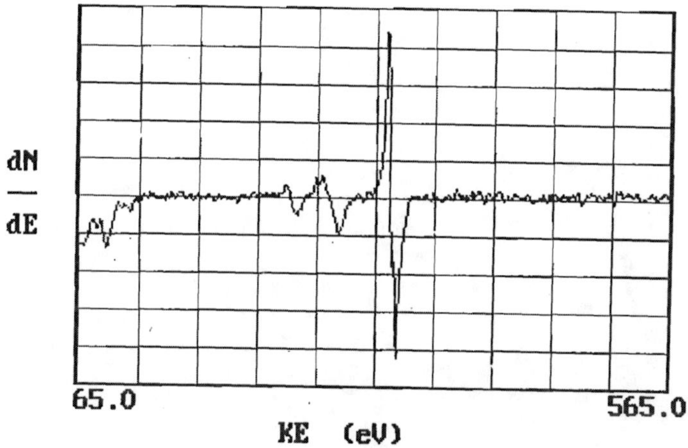

Summary

Auger Electron Spectroscopy (AES) is a surface-sensitive spectroscopic technique used for elemental analysis of surfaces: it offers

- high sensitivity (typically ca. 1% monolayer) for all elements except H and He.
- a means of monitoring surface cleanliness of samples.
- quantitative compositional analysis of the surface region of specimens, by comparison with standard samples of known composition.

In addition, the basic technique has also been adapted for use in:

1. Auger Depth Profiling: providing quantitative compositional information as a function of depth below the surface.
2. Scanning Auger Microscopy (SAM): providing spatially-resolved compositional information on heterogeneous samples.

13

Electron diffraction

Introduction

The phenomenon associated with interference processes that occur when electrons are scattered by atoms to form diffraction patterns. The wave character of electrons is shown most strikingly, and doubtless most conclusively, by the phenomena of interference. For this reason, the diffraction of electrons presents the most obvious confirmation of quantum mechanics. Because of the dependence of the diffraction pattern on the distances between the atoms, electron diffraction is also an important tool for the study of the structure of crystals and of free molecules, analogous to the use of x-rays for these purposes. See also: *x-ray crystallography; x-ray diffraction*

According to quantum theory, any particle moving with momentum mv has a wavelength $\lambda = h/mv$, where h is Planck's constant. If the particle is an electron, and its velocity is the result of having fallen through the potential difference V, this formula becomes Eq. (1), where M_0

$$\lambda = \frac{h}{(2m_0eV)^{1/2})(1 + eV/2m_0c^2)^{1/2}} \quad (1)$$

and e are the rest mass and charge of the electron, and c is the velocity of light. The last factor in the denominator represents the relativity correction, which is negligible at low voltages and amounts to only 5% at 100,000 V. For V in volts and λ in nanometers, Eq. (2) is a good approximation

$$\lambda = \left(\frac{1.5}{V}\right)^{1/2} \qquad (2)$$

to Eq. *(1)* at nonrelativistic energies. According to energy E = eV, two major techniques of structure analysis are distinguished: low-energy electron diffraction (LEED) (E ≈ 5-500 eV] and high-energy electron diffraction (HEED) [E ≃ 5-500 keV]; medium-energy electron diffraction (MEED) [E ≃ 500 eV-5 keV] is of little importance. Unlike neutrons and x-rays, electrons penetrate matter only for a very short distance before they lose energy (by inelastic scattering) or are scattered elastically (diffracted). Due to the energy loss, the wavelength of an electron changes according to Eq. (2) so that it can no longer interfere with the other incident electrons. This loss of coherence occurs in condensed matter typically after mean free paths for inelastic scattering ranging from several tenths of nanometers at low energies (E ≃ 50 eV) to several tens of nanometers at high energies, which determines the application range of LEED and HEED. See also: *Coherence; De Broglie*

Wavelength; Diffraction; Interference of waves: Mean free path; Nonrelativistic quantum theory; Quantum mechanics.

Low-Energy Electron Diffraction

LEED is used mainly for the study of the structure of single-crystal surfaces and of processes on such surfaces that are associated with changes in the lateral periodicity of the surface. A monochromatic, nearly parallel electron beam, of 10^{-4} to 10^{-3} m (4×10^{-3} to 4×10^{-2} in.) in diameter, strikes the surface, usually at normal incidence. The elastically backscattered electrons are separated from all other electrons by a retarding field and detected with a suitable movable collector or, more frequently —after acceleration to about 5 keV energy —on a hemispherical fluorescent screen with the crystal in its center. The intensity of the diffraction spots can be measured as a function of the energy $E = eV$ of the incident electrons to obtain so-called *I(V)* curves.

To obtain such results, the surface must be carefully cleaned and kept ultrahigh vacuum (UHV, pressure ≃ 10^{-16} lbf/

in.^{2} $\simeq 10^{-12}$ pascal) during the experiment and must be exposed to well-defined amounts of gases or vapors at low pressures if adsorption, condensation, or corrosion is to be studied. This requirement, as well as the fact that many surfaces are changed by the electron beam—for example, due to dissociation in the case of ionic crystals or due to desorption of adsorbed gases—limits the applicability of LEED. Nevertheless, the number of surfaces and surface processes that can have been studied is very large, and the main limitation of LEED is not caused by difficulties in obtaining LEED patterns but rather in evaluating them.

Interpretation of Patterns

The difficulties in interpreting LEED patterns are caused by the strong elastic scattering of slow electrons by atoms. As a consequence of this, the electrons observed in the LEED pattern have been scattered several times on the average, in contrast to x-ray and neutron diffraction, in which single scattering is a good approximation. Thus evaluation of the spot intensities of LEED patterns, with the additional complications arising from the elestic scattering of the surface barrier and from inelastic scattering, requires large-scale computers. As a consequence, only the structures of surfaces whose unit meshes (two-dimensional unit cells) are small have been determined by LEED. For this reason, alternative evaluation methods have been developed in which multiple-scattering effects are reduced by proper averaging procedures.

The interpretation difficulties are considerably reduced if only the lateral periodicity of the surface, but not the location of the atoms in the unit mesh, is to be determined. The determination of the lateral periodicity of the surface strucure and of the size and shape of domains with a given structure requires only the evaluaton of the geometry of the pattern, that is, of the position, size, and shape of the diffraction spots but not of their intensity, and this evaluation is relatively easy. Thus the most important application of LEED is the study of surface phenomena associated with changes in lateral periodicity. A large amount of information on lateral disorder may be extracted from spot shapes and intensities.

Surface Structure

Clean surfaces of crystals frequently have the lateral periodicity that is expected from their bulk crystal structure. The analysis of I(V) curves of such surfaces in general reveals that the distribution of the atoms normal to the surface in the topmost layers differs somewhat from that in the bulk. These relaxations, that is, displacements of atoms from their positions in the bulk, are only a few percent on densely packed surfaces. On more open surfaces they can amount to 10% or more and are frequently oscillatory. There are, however, also surfaces which show complicated superstructures (known as *reconstructed surfaces*), that is, structures whose lateral periodicity is larger than that of the substrate. Usually, such surfaces are analysed by LEED only after structural models have been proposed on the basis of other structure studies. On ideal surfaces, the decrease of the spot intensities with temperature has shown that the amplitudes of the thermal vibrations of surface atoms normal to the surface can be much larger than in the bulk. See also: *Crystal structure; Lattice vibrations*

Adsorption

Foreign atoms can interact with a clean surface in a variety of ways. The most important contribution of LEED is to the understanding of chemisorption, which precedes corrosion and, in many cases, epitaxy. Here, not only the structure of many adsorption systems, mainly of gases on metals, or metals on other metals and semiconductors, has been studied, but also the kinetics of the adsorption and desorption process as well as changes in the adsorption layer upon heating. The combination of LEED with Auger electron spectroscopy (AES) and with work-function measurements has proven particularly powerful in these studies, because such methods give the coverage and information on the locaton of the adsorbed atoms normal to the surface. Combining LEED with other complementary techniques such as ion scattering spectroscopy, electron energy loss spectroscopy, or photoelectron spectroscopy has become increasignly popular and can enable the elimination of ambiguities in the interpretation of many LEED results. See also: *Adsorption; Auger effect; Corrosion; Electron Spectroscopy; Surface and interfacial chemistry; Surface physics.*

SPLEED

This LEED with spin-polarised electrons. It gives information on the magnetic structure of surfacs and thin films due to the spin dependence of the diffraction of slow electrons in magnetic materials. This dependence is of particular importance in the spin-polarised version of low-energy electron microscopy (LEEM), which allows the magentic microstructures of surfaces and thin films to be studied.

LEEM

Low-energy diffracted electrons can be used to image the surface of a specimen if it is made the cathode in an immersion objective lens of an ultrahigh-vacuum electron microscope. Because of the strong elastic backscattering of very slow electrons, the resulting LEEM images can be recorded at video rates. This allows the study of surface processes such as phase transitions for film growth in real time. See also: *Electron microscopy*

High-Energy Electron Diffraction

HEED is used mainly for the study of the structure of thin foils, films, and small particles (thickness or diameter of 10^{-9} to 10^{-6} m or 4×10^{-8} to 4×10^{-5} in.), of molecules, and also of the surfaces of crystalline materials. A monochromatic, usually nearly parallel, electron beam with a diameter of 10^{-3} to 10^{-8} m (4×10^{-2} to 4×10^{-7} in.) is incident on the target. The forward-scattered electrons (backscattering is negligible) are detected by means of a fluorescent screen, a photoplate, or some other current-sensitive detector, usually without the inelastically scattered electrons being eliminated.

Reflection HEED

The availability of UHV instruments has made it possible to study the structure of surfaces under clean conditions by reflection HEED (RHEED). Many experiments have confirmed the theoretical expectation that RHEED has a sensitivity comparable to that of LEED. Therefore, similar to LEED, RHEED can be used for the determination of the lateral arrangement of the atoms in the topmost layers of the surface (*see Illus.),* including the structure of adsorbed layers. Although it is more convenient to deduce the periodicity of the atomic arrangement parallel to the surface from LEED patterns than from RHEED patterns, LEED frequently becomes inapplicable when the surface is rough. This usually occurs in the later stages of corrosion or in precipitation, for example, of silicon carbide on silicon, when small crystals grow on the surface. In such investigations RHEED is far superior to LEED because the fast electrons can penetrate the asperities and produce a transmission HEED pattern. RHEED has also become important for thin-film growth monitoring via the specular beam intensity oscillations caused by monolayer-by-monolayer growth.

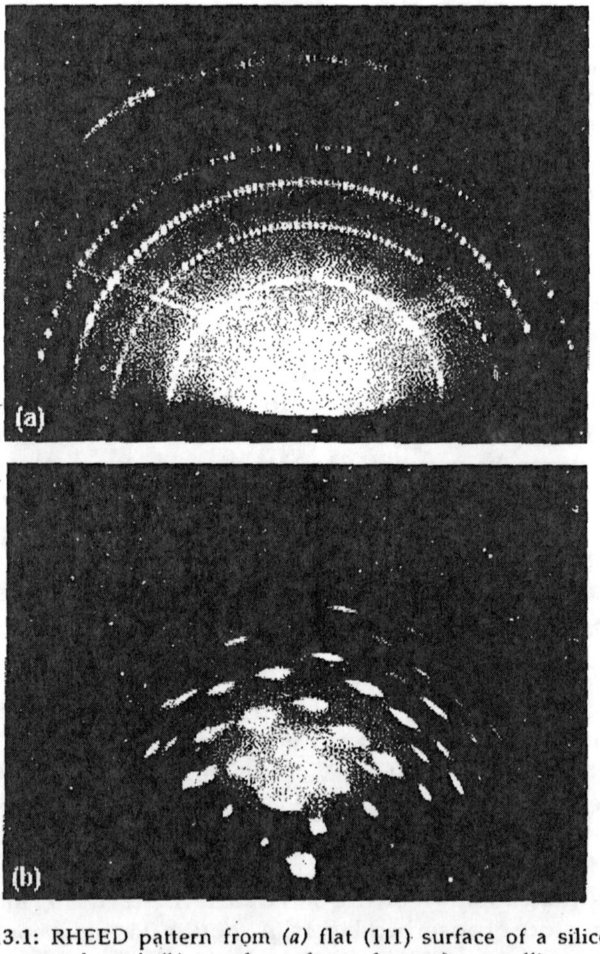

Fig. 13.1: RHEED pattern from *(a)* flat (111) surface of a silicon single crystal, and *(b)* rough surface of a polycrystalline evaporated film of calcium fluoride (CaF_2) with strong fiber texture, from which no LEED patter can be obtained. & plusvsp5;

Like LEED, RHEED gives little informtion on the chemical nature of the atoms that produce the diffraction pattern. Therefore, RHEED has been combined with analytical tools, such as Auger electron spectroscopy or x-ray emission spectroscopy. A second limitation inherent to both RHEED and LEED is the difficulty of determining the atomic positions normal to the surface. This is a consequence of dynamical effects and of absorption. Nevertheless, both techniques have given valuable information on corrosion, precipitation, adsorption, condensation, film growth, and other surface and interface phenomena that could not be obtained by other methods.

Scanning HEED

In scanning HEED (SHEED) the diffracted electrons are not recorded on photographic film but are directly measured electronically with sensitive detectors. By moving the detector across the diffraction pattern or by deflecting the diffracted electrons across a stationary detector (scanning), the intensity distribution in the diffraction pattern can be displayed quantitatively on an XY recorder. If an energy filter is put in front of the detector, the inelastically scattered electrons, which are usually not taken into account in the quantitative intensity evaluation, may be filtered out so that only the elastically scattered electrons are measured. This technique is particularly useful in transmission through polycrystalline samples, which produce a ring pattern (Debye Scherrer diagram). It has also been used in single-crystal samples producing a spot pattern (Laue diagram) and even in reflection diffraction from surfaces. The main application of SHEED is in the study of processes which are accompanied by changes of the intensity distribution, such as the growth of thin films and annealing and corrosion processes.

Transmission HEED

The technological importance of thin film and interface devices has led to an upsurge of thin film growth studies by conventional transmission HEED (THEED), usually combined with transmission microscopy. Information obtained this way has been mainly on the orientation of the crystallites composing the film.

Diffraction electron microscopy

In transmission electron microscopy of crystalline specimens the main contrast mechanism is the diffraction contrast. To understand it, knowledge of the diffraction pattern and of diffraction theory is necessary. High-voltage electron microscopes (with electron energies of about 1 MeV) have become available. As a consequence, samples with thickness of the order 10^{-6} m (4×10^{-5} in.) can now be studied, as compared to the former upper thickness limit of about 10^{-7}mm (4×10^{-6} in.). This has significance for metallurgy and the physics of metals and semiconductors. It is now possible to study the structure, distribution, and behaviour of imperfections in metals, alloys, and semiconductors in bulk with little influence from the boundaries of the material. Other electron-microscope techniques that rely heavily on diffraction are the use of illumination systems producing tilted or conical illumination of the specimen and of objective apertures which transmit only diffracted electrons. With these techniques it has become possible to produce, by diffraction contrast, images of crystal planes 1.5×10^{-10}m (6×10^{-9} in.) apart. RHEED beams can be used for reflection electron microscopy (REM), though with considerable foreshortening due to the grazing incidence on the surface which is necessary for surface sensitivity and intensity returns.

Diffraction in Gases and Liquids

Electron diffraction in gases and liquids is similar in principle to that in solids; the differences arise from the lack in gases and liquids of any highly regular arrangement of the component atoms. In gases the low density makes it possible to study diffraction by individual atoms and molecules. The results obtained from monatomic gases represent the density of electronic charge in the atom as a function of the distance from the nucleus . The results from gaseous polyatomic molecules represent the equilibrium distances between the atomic nuclei and the average amplitudes of vibration associated with these distances. Liquids have been studies much less thoroughly, both in theory and in practice, than have gases.

It should be rememered that the structures of molecules in the gaseous state may be different from those in the liquid or solid states; for example, hydrofluoric acid in the gas form an $(HF)_6$ ring, while in solution or liquid it forms and infinite chain (— FHFHFHFHFHFHF—); biphenyl in the gas is twisted with conformational equilibria, whereas in the solid state it is coplanar in one frozen conformer.

Applications

Typical questions of molecular structure studied by electron diffraction include those of configuration and size in many molecules, with special interest in the variation of chemical bond distances in related molecules (such as the 6.8 picometers or 2.7×10^{-10} in. decrease in C_F distance in the series CH_3F, CH_2F_2, CHF_3, CF_4), the variation in the angles between chemical bonds, the distinction between geometric isomers, the degree of restricted rotation around chemical bonds, and in general the relations between the geometry of molecules and their energy and chemical behaviour. See also *Chemical bonding; Molecular isomerism; Molecular structure and spectra; Strctural Chemistry*

The applicaiton of electron diffraction in liquids has been restricted because of the less precise information obtainble from molecules which lie in close contact but in irregular arrangements.

The liquids studies include certain metals (mercury, tin, tin-bismuth, and tin-aluminum alloys), paraffin oils, and thin layers of water. For monatomic molecules in the liquid state the distances between nearest neighbours are fairly uniform, whereas the separation between more distant neighbours varies considerably. Evaluation of the closest distance of approach can be made to about 10 pm (4×10^{-10} in.); major developments in the theory are still required.

The most interesting application of electron diffraction to a liquid or amorphous condensed phase has been in the examination of the surface layers on polished solids. The diffuse diffraction patterns often obtained may be the result of an amorphous arrangement in the surface or of the poor resolving power of very tiny crystalline particles. It is probable that both these states are produced in the polishing of different solids.

Theory and Techniques

Structural information about gaseous molecules is obtained by háving a fine beam of electrons pass through the gas and strike a photographic plate. The electrons in the beam interact with the charaged particles in the atoms (electrons and nuclei) and are bent away from the original direction through varying angles, as registered in the pattern on the plate. The observed variation in the number of scattered electrons with increasing anagle is interpreted as an interference effect; that is, electron scattering can be described in the language of diffracted waves for which the resultant is the sum of the component wavelets whose amplitudes and phases are influenced by the wavelength of the incident radiation and the relative positions of the scattering centers. The equivalent wavelength of the electrons, λ, is determined by their energy and is given by Eq. (1) or (2). The values of λ commonly used in diffraction by gases are in the range 7- 5 $\times 10^{-12}$ m ($3^{-2} \times 10^{-10}$ in).

The intensity *I* of electrons scattered by a spherically symmetrical atom is computed by the Schrodinger wave equation; the result is given by Eq. (3)

$$I = k\,|f|^2 \qquad (3)$$

where *k* is a constant and *f* is a complex function called the atomic scattering factor. If the atomic number Z is not too large, Eq. *(4)* holds,

$$|f|^2 = [\,2\,(Z - F)/\,as^2]^2 \qquad (4)$$

where *a* is the Bohr radius and *F* is given by Eq. *(5)*

$$F = 4\pi \int_0^\infty r^2\,\rho(r)\,[(\sin sr)/sr]dr \qquad (5)$$

Here $\rho(r)$ is density of electronic charge at distance *r* from the nucleus, $s = 4\pi\,(\sin\theta)/\lambda$ with 2θ equal to the angle between the scattered electron and the original beam, and λ is the electron wavelength. Experimental observations on I as a function of *s* in monatomic gases lead to the determination of $\rho(r)$

The intensity of electrons scattered by a collection of independence molecules having all possible orientations is given by Eq. (6), where the

$$I(s) = k \sum_{i} [|f_i|^2 + (2/as^2)^2 S_i] + \sum_{i}\sum_{j} f_i f_j \int_0^{\infty} P_{ij}(r) [(\sin sr)/sr] dr \quad (6)$$

summations are taken over all the atoms in the molecule and $(2as^2)^2 S_i$ represents the intensity of inelastic scattering by the *i*th atom. The double summation has a term for each pair of atoms *i* and *j*. The relative probability of finding the distance between the atom *i* and *j* at various values is represented by $P_{ij}(r)$. In principle the use of the observed intensity to determine P_{ij} for each pair of atoms in the molecule constitutes a structure determination for the molecule. The expression is simpler when the atomic motions are nearly harmonic, as in the case of CCl_4. For this molecule the double summation has only two distinct terms. The first is given by expression (7); the second for Cl—Cl is similar.

$$8 f_C f_{Cl} \exp(-l^2_{CCl} s^2)(\sin sr_{CCl})/sr_{CCl} \quad (7)$$

The four parameters which describe the structure are the equilibrium distances, r_{CCl}, and r_{ClCl}' and the average amplitudes of vibration, l_{CCl} and l_{ClCl}. Molecular parameters such as these can be determined with high precision when a proper treatment of the scattering factors is included, and when accurate experimental intensities are obtained.

Special equipment is required to meet the conditions assumed in the scattering theory. The electron beam is accelerated with a steady voltage between 30,000 and 70,000 V, which should be constant within about 0.01%. The beam is focused by electrostatic and magnetic lenses so that it has a diameter of no more than 0.1 mm (0.004 in.) at the photographic plate. The whole path of the beam is enclosed in a high-vacuum chamber (pressure of about 10^{-5} mm Hg or 10^{-3} Pa or 10^{-7} ibf/in.) so that no appreciable electron scattering will occur in the residual air. The gas specimen is introduced in a fine jet so that the volume in which the electrons meet the gas is no more than 1.0 mm (0.04 in.) in diameter; high-speed pumping is required to remove the gas as rapidly as possible. In front of the photographic plate a rotating sector is mounted to modify the intensity of electrons reaching the plate so that the normally rapid decrease of intensity with increasing angle of scattering is levelled off in a known way and the emulsion is able to register the incident electrons over a wide range of angle.

Interpretation of Results

The pattern observed is a set of light and dark circular bands whose spacings and relative intensities depend on the composition and structure of the specimen molecules. The pattern is scanned by a recording microphotometer which yields (after calibration of the photographic emulsion) a tracing or a digital output of the experimental data of I as a function of s. These data are interpreted with the aid of a Fourier transform of the intensity expression of Eq. (6) to give the P_{ij} functions for the pairs of atoms. See also: *Fourier series and transforms*

The relatively high precision in determining the distances in gas molecules has been achieved since about 1950 with the development of new instrumentation and the application of high-speed computing methods. The relative intensities of scattered electrons registered on the photograhic plate can be measured to better than 0.1%; the time saved by the use of electronic computing methods in interpreting the data permits a more rigid application of the criteria for satisfactory agreement between experiment and theory.

Over a thousand gases have been studied by the methods giving internuclear distances within 0.4 pm (1.6×10^{-11} in.) In favourable cases precisions exceeding 0.1 pm (4×10^{-12} in.) have been obtained; when these are compared with spectroscopic results, good agreement is found if allowance is made for the difference in the nature of the distances measured. The molecular structures of some 500 gases have been reported by earlier electron diffraction procedures with uncertainties 10 times larger or more.

The applicability of the gas diffraction method alone is limited to simple molecules. Only when the three-dimensional structure can be derived uniquely from a one-dimensional spectrum of internuclear distances blurred by thermal motion is a complete structure determination possible from electron diffraction data alone; light atoms in the presence of heavy ones in the same molecule are less precisely located; distances as close as 3 pm (1.2×10^{-10} in.) can barely be resolved. The range of the method is greatly increased, however, when some structural features in the molecule can be assumed from the results of other methods of investigation. For example, if a six-membered ring of atoms is known to have trigonal symmetry, the number of structural parameters is decreased from 12 to 2.

14

Reflection High Energy Electron Diffraction (RHEED)

Introduction

Low Energy Electron Diffraction (LEED) utilises the inherent surface sensitivity associated with low energy electrons in order to sample the surface structure. As the primary electron energy is increased not only does the surface specificity decrease but two other effects are particularly noticeable.

1. forward scattering becomes much more important (as opposed to the backward scattering observed in LEED)
2. the scattering angle (measured from the incident beam direction) tends towards 180 degrees for back-scattering and 0 degrees for forward scattering.

In order to extract surface structural information from the diffraction of high energy electrons, therefore, the technique has to be adapted and the easiest way of doing this is to use a reflection geometry in which the electron beam is incident at a very grazing angle - it is then known as Reflection High Energy Electron Diffraction (RHEED).

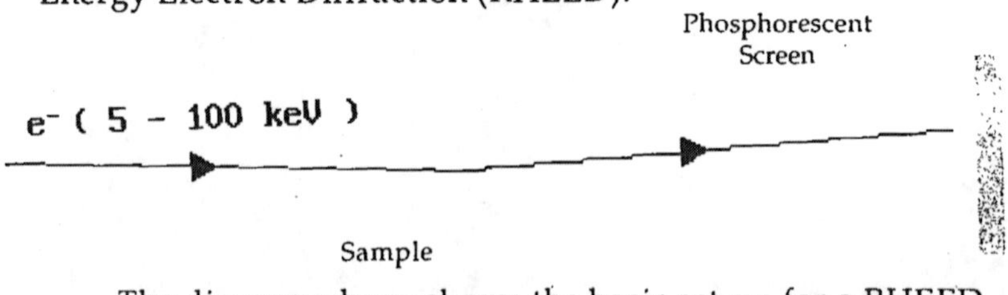

The diagram above shows the basic set-up for a RHEED experiment, with the sample viewed edge-on. In practice the display screen is usually a phosphor coating on the inside of a vacuum window (viewport) and the diffraction pattern can be viewed and recorded from the atomspheric side of the window. The small scattering angles involved are compensated for by using relatively large sample/screen distances.

The sample can be rotated about its normal axis so that the electron beam is incident along specific crystallographic directions on the surface.

In order to understand the diffraction process we need to consider how the eletron beam can interact with the regular array of surface atoms in this experimental geometry. It is worth noting, however, that the use of glancing incidence ensures that, despite the high energy of the electrons, the component of the electron momentum perpendicular to the surface is small. Under these conditions an electron may travel a substantial distance through the solid (in accord with the much longer mean free path of such high energy electrons) without penetrating far into the solid. The technique, consequently, remains surface sensitive.

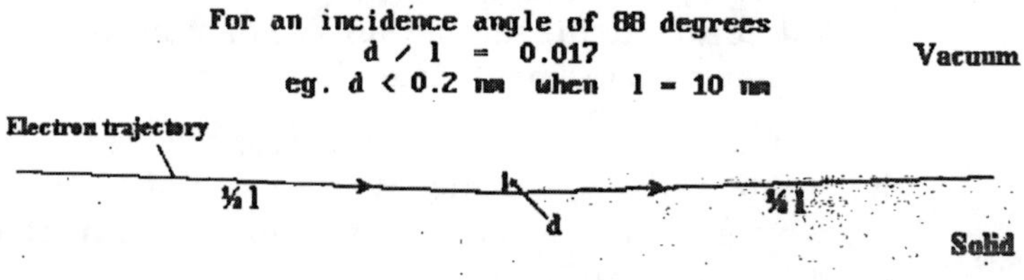

Now consider the plan view of a surface illustrated below in which we concentrate attention on just one row of atoms (shown shaded in pale blue) running in a direction perpendicular to the incident electron beam (incident from the left).

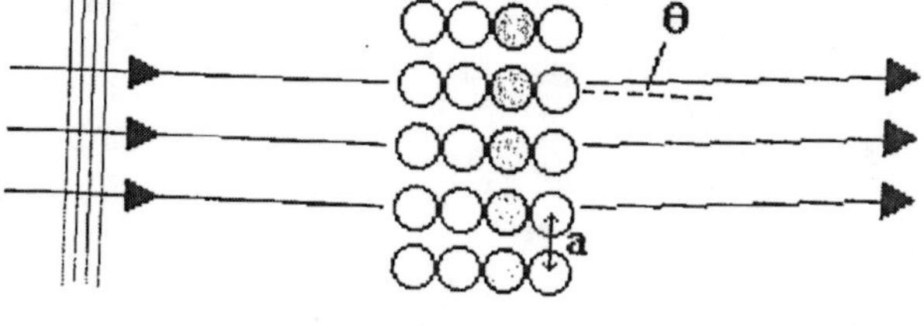

In addition to the change in momentum of the electron perpendicular to the surface, which leads to the apparent reflection, the diffraction process may also lead to a change in momentum parallel to the surface, which leads to the deflection by an angle θ when looked at in plan view. Constructive interference occurs when the path difference between adjacent scattered "rays" ($a \sin \theta$) is an integral number of wavelengths (i.e. the same basic condition as for LEED). This gives rise to a set of diffracted beams at various angles on either side of the straight through (specularly reflected) beam.

What, if any, advantages does RHEED offer over LEED?

In terms of the quality of the diffraction pattern absolutely none ! moreover, diffraction patterns have to be observed for at least two sample alignments with respect to the incident beam in order to determine the surface unit cell. However,...

1. The geometry of the experiment allows much better access to the sample during observation of the diffraction pattern. This is particularly important if it is desired to make observations of the surface structure during growth of a surface film by evaporation from sources located normal to the sample surface or simultaneous with other measurements (e.g. AES, XPS).

2. Experiments have shown that it is possible to monitor the atomic layer-by-atomic layer growth of epitaxial films by monitoring the intensity of the diffracted beams in the RHEED pattern.

By using RHEED it is therefore possible to measure, and hence also to control, atomic layer growth rates in Molecular Beam Epitaxy (MBE) growth of electronic device structures - this is by far and away the most important application of the technique.

15

Basic Concepts in Surface Imaging and Localised Spectroscopy

Introduction

Most surface spectroscopic techniques involve probing the surface by exposing it to a flux of "particles" ($h\nu$, e^-, A^+ . . . .) and simultaneously monitoring the response to this stimulation by, for example, measuring the energy distribution of emitted electrons. In their most basic form, these techniques collect information from a relatively large area of surface (~mm^2). In most cases, however, there are variations of these techniques which permit either:

1. information to be collected only from a specific region of the surface - that is.... **Localised Spectroscopy**

   or

2. the spatial distribution of a property (eg. simple topography or elemental concentration) to be mapped - that is..

   Surface Imaging (Surface Microscopy)

The requirement in both cases is for **spatial localisation** of the spectroscopic technique. This may be achieved in one of two ways.

A. Localisation of the Probe

Spatial localisation of a spectroscopic technique can be achieved by focusing the probe so that it interacts with only a small region of the surface.

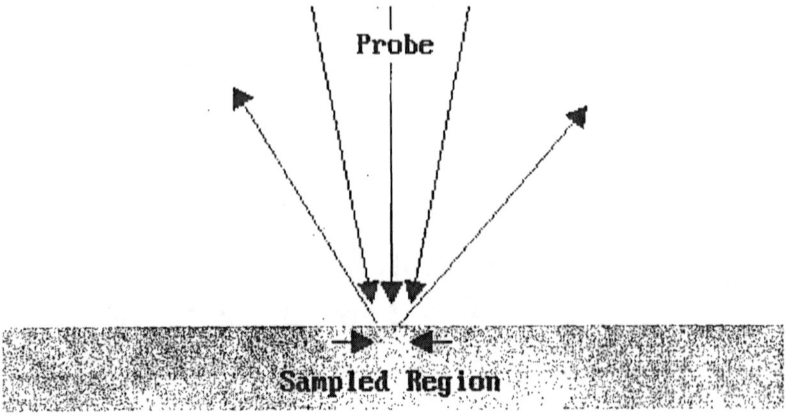

The ease with which this can be accomplished and the spatial resolution (degree of localisation) attainable depends primarily upon the nature of the probe - charged particles (e.g. electrons and ions) are easily focused; neutral particles (e.g. x-ray photons and neutral atoms) are not readily focused.

With this localisation of the probe into a small area, it is clearly advantageous to collect as many as possible of the outgoing particles whose detection completes the spectroscopic process in order to maximise the signal-to-noise.

In the diagram above, the probe is stationary and incident upon a particular small region of the surface in a manner suitable for localised surface spectroscopy. In order to carry out surface imaging or mapping, either

- the probe must be scanned *(rastered)* across the surface, or
- the surface must be moved under the probe beam.

B. Spatially-localised Detection

The alternative approach is to allow a large area of the sample to be exposed to the probe but only to collect signal (i.e. emitted particles) from a small region of the sample. Thus, in this instance, the spatial localisation is occurring on the detection side of the technique.

In principle, this is easy - it simply requires an appropriate lens system to collect as much emission as possible from the desired part of the surface and focus it onto a detector as illustrated in a highly schematic fashion below.

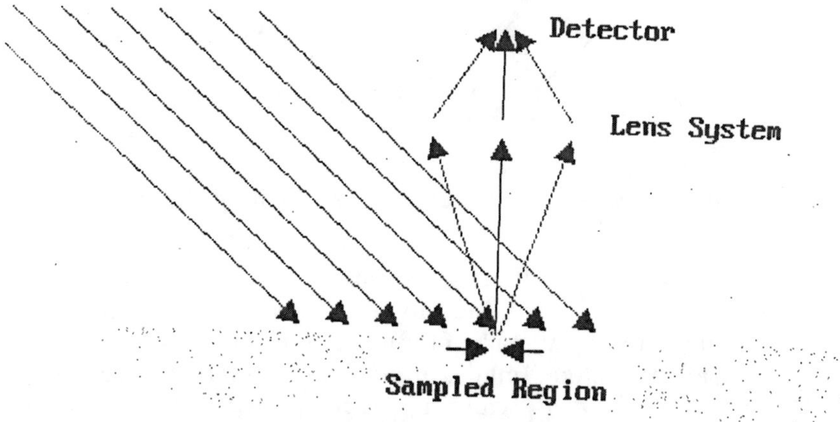

In practice, this approach has only really been successfully employed in surface analysis when the emitted particles are electrons.

In order to get an image of the surface using a single detector it is necessary to either

1. scan the sample position underneath the detection system.
2. make use of additional scanning plates in the electon-optical focusing system.

An alternative approach to obtaining images using this same basic arrangement involves employing a linear array or two-dimensional array of detecting elements. In this type of system, the emission from different parts of the surface is effectively focussed onto different parts of the array detector as schematically illustrated below.

Some additional scanning mechanism is still required to build-up an image or map of anything other than the directly imaged region of surface, but the use of multiple (array) detectors

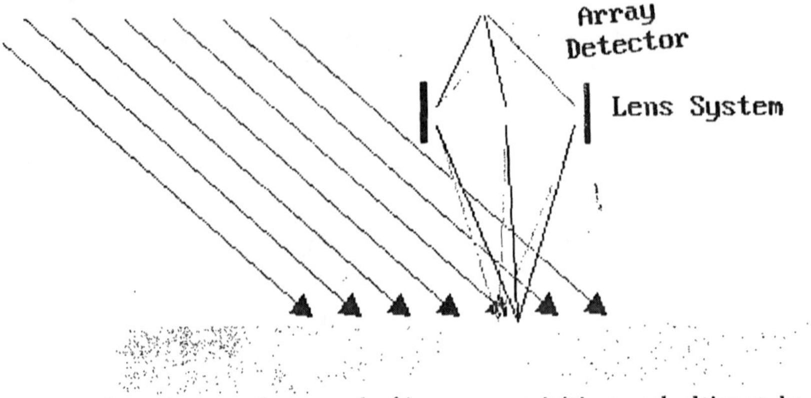

greatly enhances the speed of image acquisition and ultimately the image quality.

In summary, spatial localisation of a surface spectroscopic technique may be obtained by either focusing the probe or imaging only a small part of the sample surface onto a detector using a lens system. Images and compositional maps are then usually obtained by scanning the focused probe in the first case, and either translating the sample or employing sophisticated optical/detector systems in the latter case.

SIMS - Imaging and Depth Profiling

The basic ideas behind the SIMS (Secondary Ion Mass Spectrometry) technique have already been discussed in previous chapter.

Since the technique utilises a beam of atomic ions (i.e. charged particles) as the probe, it is a relatively easy matter to focus the incident beam and then to scan it across the surface to give an imaging technique.

A. Surface Imaging using SIMS

If the aim of the measurement is to obtain compositional images of the surface formed from the secondary ion spectrum with minimum possible damage to the surface, then the main problem is to ensure that sufficient signal is obtained at the desired spatial resolution whilst minimising the ion flux incident on any part of the surface.

This is most easily achieved by switching from the traditional instrumental approach of using continuous-flux ion guns and quadrupole mass spectrometer detectors, to using pulsed ion sources and time-of-flight (TOF) mass spectrometers. The TOF mass spectrometers are a much more efficient way of acquiring spectral data, and also provide good resolution and sensitivity up to very high masses. Using such instruments, SIMS images with a spatial resolution of better than 50 nm are obtainable.

B. SIMS Depth Profiling

The aim of depth profiling is to obtain information on the variation of composition with depth below the initial surface - such information is obviously particularly useful for the analysis of layered structures such as those produced in the semiconductor industry.

Since the SIMS technique itself relies upon the removal of atoms from the surface, it is by its very nature a destructive technique, but also, ideally suited for depth profiling applications. Thus a depth profile of a sample may be obtained simply by recording sequential SIMS spectra as the surface is gradually eroded away by the incidention beam probe. A plot of the intensity of a given mass signal as a function of time, is a direct reflection of the variation of its abundance/concentration with depth below the surface.

One of the main advantages that SIMS offers over other depth profiling techniques (e.g. Auger depth profiling) is its sensitivity to very low concentrations (~ppm) of elements - again this is particularly important in the semiconductor industry where dopants are often present at very low concentrations.

The *depth resolution* achievable (e.g. the ability to discriminate between atoms is adjacent thin layers) is dependent upon a number of factors which include:

1. the uniformity of etching by the incident ion beam.
2. the absolute depth below the original surface to which etching has already been carried out.
3. the nature of the ion beam utilised (i.e. the mass & energy of the ions).

as well as effects related to the physics of the sputtering process itself (e.g. ion-impact induced burial).

Auger Depth Profiling

As described in earlier, Auger Spectroscopy is a surface sensitive spectroscopic technique yielding compositional information.

In its basic form it provides compositional information on a relatively large area (~ 1 mm^2) of surface using a broad-focused electron beam probe. In this manner, sufficient signal can be readily obtained whilst keeping the incident electron flux low, and thus avoiding potential electron-induced modifications of the surface. As a consequence the technique is non-destructive when used in this manner.

To obtain information about the variation of composition with depth below the surface of a sample, it is necessary to gradually remove material from the surface region being analysed, whilst continuing to monitor and record the Auger spectra.

This controlled surface etching of the analysed region can be accomplished by simultaneously exposing the surface to an ion flux which leads to sputtering (i.e. removal) of the surface atoms.

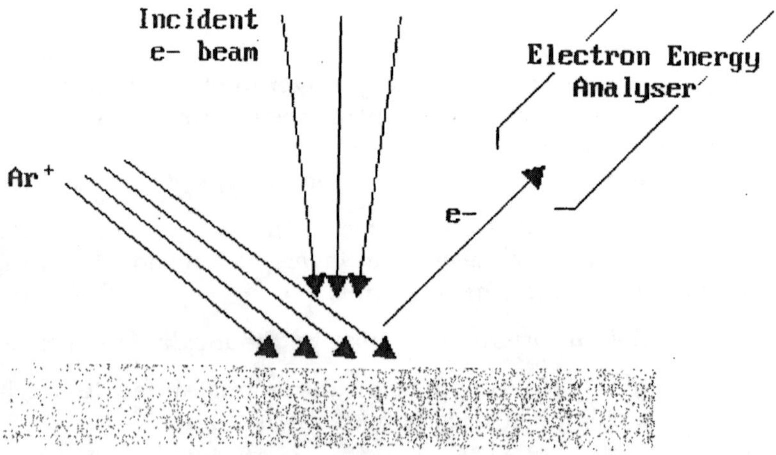

For example, suppose there is a buried layer of a different composition several nanometres below the sample surface.

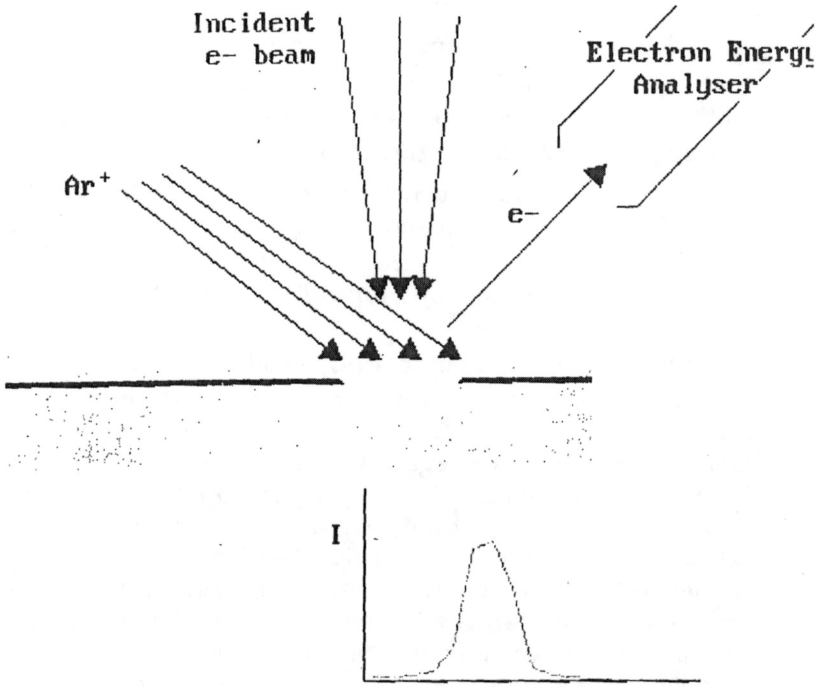

As the ion beam etches away material from the surface, the Auger signals corresponding to the elements present in this layer will rise and then decrease again.

The diagram above shows the variation of the Auger signal intensity one might expect from such a system for an element that is only present in the buried layer and not in the rest of the solid.

In summary, by collecting Auger spectra as the sample is simultaneously subjected to etching by ion bombardment, it is possible to obtain information on the variation of composition with depth below the surface—this technique is known by the name of *Auger Depth Profiling*.

Electron Microscopy - SEM & SAM

The two forms of electron microscopy which are commonly used to provide surface information are

Secondary Electron Microscopy (SEM)

- which provides a direct image of the topographical nature of the surface from all the emitted secondary electrons.

Scanning Auger Microscopy (SAM)

- which provides compositional maps of a surface by forming an image from the Auger electrons emitted by a particular element.

Both tecniques employ focusing of the probe beam (a beam of high energy electrons, typically 10 - 50 keV in energy) to obtain spatial localisation.

A. Secondary Electron Microscopy (SEM)

As the primary electron beam is scanned across the surface, electrons of a wide range of energies will be emitted from the surface in the region where the beam is incident. These electrons will include backscattered primary electrons and Auger electrons, but the vast majority will be *secondary electrons* formed in multiple inelastic scattering processes (these are the electrons that contribute to the background and are completely ignored in Auger spectroscopy). The secondary electron current reaching the detector is recorded and the microscope image consists of a "plot" of this current, I, against probe position on the surface. The contrast in the micrograph arises from several mechanisms, but first and foremost from variations in the surface topography. Consequently, the secondary electron micrograph is virtually a direct image of the real surface structure.

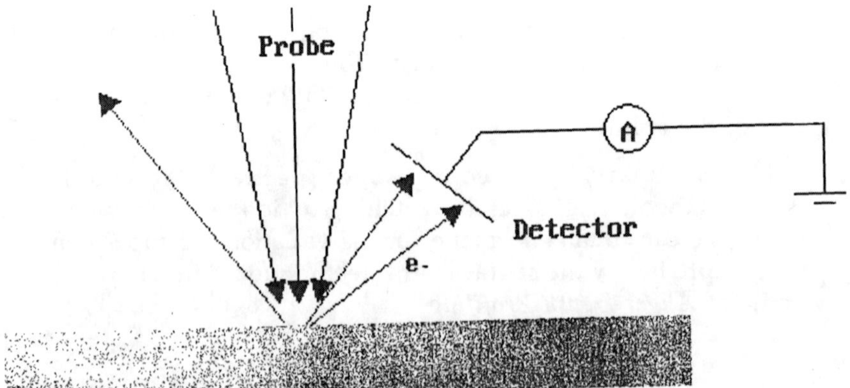

The attainable resolution of the technique is limited by the minimum spot size that can be obtained with the incident electron beam, and ultimately by the scattering of this beam as it interacts with the substrate. With modern instruments, a resolution of better than 5 nm is achievable. This is more than adequate for imaging semiconductor device structures, for example, but insufficient to enable many supported metal catalysts to be studied in any detail.

B. Scanning Auger Microscopy (SAM)

The incident primary electrons cause ionization of atoms within the region illuminated by the focused beam. Subsequent relaxation of the ionised atoms leads to the emission of Auger electrons characteristic of the elements present in this part of the sample surface.

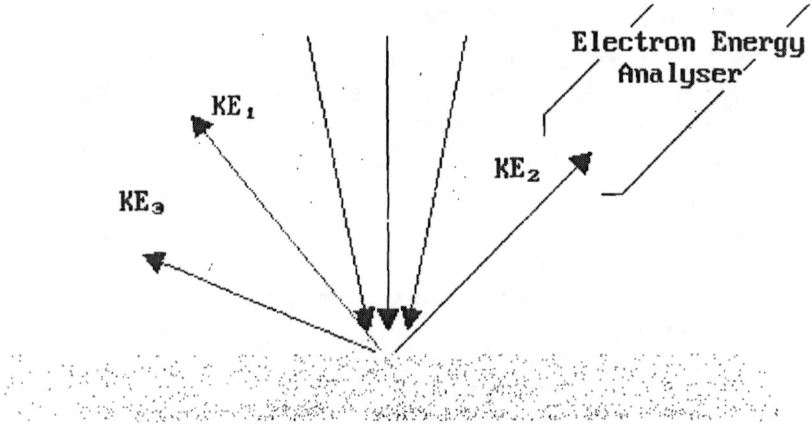

As with SEM, the attainable resolution is again ultimately limited by the the incident beam characteristics. More significantly, however, the resolution is also limited by the need to acquire sufficient Auger signal to form a respectable image within a reasonable time period, and for this reason the instrumental resolution achievable rarely approaches 20 nm.

Imaging XPS

The combination of the features of X-ray photoelectron spectroscopy (in particular, quantitative surface elemental analysis and chemical state information) together with spatial localisation is a particularly desirable option in surface analysis. However, whilst much progress has been made in developing the technique of imaging XPS, there is still a considerable research effort being devoted to improving the available spatial resolution beyond that which is currently available.

Different manufacturers of imaging XPS systems have adopted different strategies for obtaining spatial localisation - including all of those mentioned earlier. Specificially, these include

1. Localisation of the probe, by focusing the incident x-rays.
2. Localisation by selected (limited) area analysis.
3. Use of array detectors, with associated imaging optics.

1. Imaging XPS by X-ray Focussing

Until very recently this was not a popular option, since x-rays are rather difficult to focus (in fact x-ray lens design is a science in its own right and alternative focusing techniques employing especially-bent diffracting crystals have to be employed).

Nevertheless, with recent improvements in technology, an x-ray spot size of better than 10 μm has been achieved using this approach; images may then be otained *either* by scanning the sample under the focused x-ray beam, or by scanning the microfocused x-ray beam.

2. Limited Area Analysis

The most common approach to localisation of analysis, is to restrict the area of the sample surface from which photoelectrons are analysed using a combination of lenses and apertures in the design of the electron energy analyser.

The main problem with this approach is that as the sampled area is reduced, so is the collected signal - consequently, there is a direct trade-off between spatial resolution and data collection time.

The practically achievable spatial resolution is rarely better than 100 μm.

Imaging of the sample surface may then be achieved by either:

1. translating the sample position under the electron energy analyser, so that the analysed region is moved across the surface.
2. incorporating electrostatic deflection plates within the electron optics to move the region from which electrons are collected across the sample surface.

One advantage of this technique is its relative simplicity (and hence relatively low cost) - but this is reflected in the poor performance!

3. Array Detectors and Imaging Optics

There are a number of innovative designs for imaging spectrometers which combine array (i.e. multiple-segment) detectors with sophisticated imaging optics to obtain electron-energy resolved images at much faster acquisition rates. The spatial resolution achievable using this approach is also higher than that for either of the other tecniques meantioned - state-of-the-art instruments give better than 10 μm resolution.

In summary, the spatial resolution currently achievable with imaging XPS instruments limits their potential applications - nevertheless there are many areas of materials science where the information obtainable is incredibly useful and the relatively poor spatial resolution (compared, for example, with electron microscopic techniques such as SAM) is more than offset by the benefit of concurrent chemical state definition.

16

Adsorption Potential Variation on a Real Surface

Introduction

It is important to realise that the adsorption potential of even such a simple adsorbate as a monatomic gas, will vary with position on the surface of even such a surface as an ideal graphite adsorbent. If we define a quantity ${}_mU_o$ given by

$$ {}_mU_o = U_o + E_o^{vib} \quad (1) $$

then this could be evaluated by calculating all the separate interactions between the adsorbed atom at a particular position at distance *d* from each of the carbon atoms in the substrate, and then summing the contributions for each individual interaction. Thus the interaction of the gas molecule (i) with each of the atoms (j) of the adsorbent could be written

$$ u_{ij} = C_1 d_{ij}^{-6} + C_2 d_{ij}^{-8} + C_3\, d_{ij}^{-10} + R\, d_{ij}^{-12} \quad (2) $$

where C_1, C_2 and C_3 are the dipole-dipole, dipole-quadrupole and quadrupole-quadrupole coefficients, respectively; *R* is the Born repulsion constant. The total interaction will therefore be

$$ {}_mU_O = \sum_j u_{ij} = C_1 \sum_j d_{ij}^{-6} + C_2 \sum_j d_{ij}^{-8} + C_3 \sum_j d_{ij}^{-10} + R \sum_j d_{ij}^{-12} \quad (3) $$

Avgul *et al* have calculated values of ${}_mU_o$ for neon, argon and krypton at three different positions on a graphite surface, and these are illustrated in Fig. 16.1. Their results are tabulated in Table 16.1.

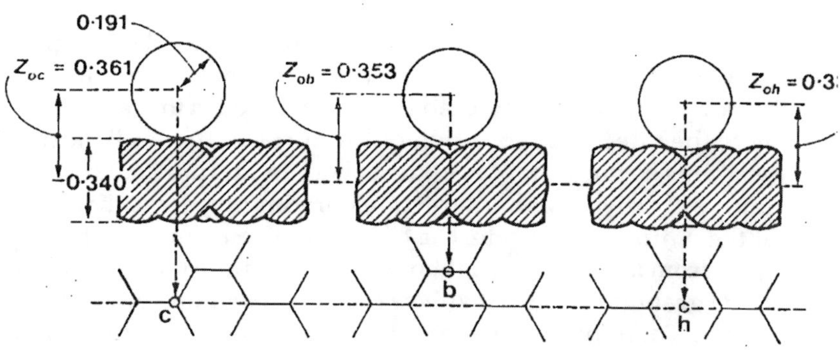

Fig. 1: Adsorbed argon atoms at different positions on a graphite substrate: all distances in nm.

Table 16.1

Calculated values of $_{m}U_{o}$/kJ mol^{-1} for inert gas molecules at three positions on a graphite surface

	c *above C atom*	*b* *between 2 C atoms*	*h* *centre of C ring*	*Average*
Neon	3.22	3.51	5.64	3.81
Argon	8.16	8.66	11.05	9.29
Krypton	10.84	11.63	14.52	12.34

This table clearly illustrates two particularly significant facts. Firstly the increase in $_{m}U_{o}$ with increasing adsorbate molecular size, and secondly the appreciable variation in $_{m}U_{o}$ with position. Whether these variations would be large enough to hinder the mobility of an adsorbed inert gas atom would be dependent on the relationship of the variation in $_{m}U_{o}$ to the energy associated with one translational degree of freedom, ½kT. At 77.5 K, the latter corresponds to 0.16 kJ mol^{-1}, and the variation in $_{m}U_{o}$ from the table is 1.42 KJ mol^{-1} in the case of neon and 3.68 kJ mol^{-1} for krypton. Direct comparison however takes no account of the magnitude of E_0^{vib}: we will return to this point again later.

In the case we have just considered there is a high degree of symmetry in the surface, and the positional variations can be easily represented in the table. Also neglected in the calculation were the effects of relaxation of the surface from bulk lattice positions as the surface is approached. In this particular case, since only dispersion interactions are involved and because of the nature of graphite, the bulk of the contributions to ${}_mU_o$ come from the nearest neighbours, and also probably relaxation of the graphite surface involves the movement of whole lattice planes. Therefore the overall picture described in Table (16.1) will be substaintially correct.

The calculation also excluded the possibility of secondary relaxation of the surface brought about by the adsorption of the inert gas atom itself. In this case again the effect is unlikely to be significant since ${}_mU_o$ is much smaller than the bond energy holding the graphite plane together. However, if the forces holding the solid together were comparable with the adsorption forces, then the effect might be significant.

Burton and Jura considered the effect of relaxation on the value of ${}_mU_o$ for argon, neon and krypton adsorbed on the surface of solid argon, and their results are shown in Table 16.2. The changes are clearly significant in terms of the value of ${}_mU_o$, and there are also positional changes for the adsoprtion of an additional argon atom. The nearest neighbours to the adsorbed atom relax towards the bulk 0.6% of the lattice spacing from their normal surface positions and outwards tangentially to the surface by 0.3%.

Table 16.2

Values of ${}_mU_o$/kJ mol^{-1} for argon, neon and krypton adsorbed on the (100) face of solid argon, with and without allowance for relaxation.

	Argon	*Neon*	*Krypton*
Without relaxation	5.67	2.48	6.60
With relaxation	5.72	2.87	7.04

Adsorption potential maps: solid inert gases

The problem of calculating adsorption potentials before about 1960, was largely the amount of tedious arithmetic that had to be done. The advent of the modern computer has changed the situation drastically. Although the work involved in writing an adequate programme and debugging it is not inconsiderable, the aditional work involved in obtaining answers at a very large number of points on the surface is no longer a problem. Therefore sufficiently detailed results can be obtained to permit their presentation in the form of maps of the surface, where the x- and y-directions represent distances on the surface, and the z-direction adsorption potential contours. Thus the difficulty now lies in the formulation of an adequate physical and mathematical model on which to base the calculations, and all the difficulties that this presents have by no means been resolved. To illustrate the methods and the problems, even with a comparatively simple case, we will look at the results obtained by Ricca *et al* for the adsorption of an inert gas molecule on a solid inert gas substrate.

The calculations were based on the assumption that the total interaction energy can be evaluated by adding the pair-wise values of the interaction between the adsorbing atom and each atom in the substrate. The individual values were calculated using the Lennard-Jones '6-12' potential equation

$$U(r) = \varepsilon_0 \left[\frac{r_0}{r} \right]^{12} - 2 \left[\frac{r_0}{r} \right]^{6} \qquad (4)$$

where ε_0 is the depth of the energy mininum and r_0 is the value of the separation, r, at this point. An example of their results is shown in Fig. 16.2 as a contour map of the potential energy surface, showing that the adsorption site is at the mid-cell position *, the minimum translational barrier being over the saddle point to a neighbouring site. In order to estimate the degree of localisation, or state of mobility, of the adsorbed atoms of the surface, Ricca stated that simple comparison of mean translational energy levels with translational barrier heights was inadequate. The exact nature of the surface can be better visualised by the section illustrated in Fig. 16.3 which

* The caculations assumed ideal bulk crystal positions throughout, that is, no relaxation due to the presence of the surface, and no secondary relaxation as adsorption occurs.

represents the vertical plane through Fig. 16.2 at the position represented by a horizontal line joining A-A and passing through S, halfway up the contour map. It should be noted that the equilibrium adsorption distance from the surface plane is a positional variable.

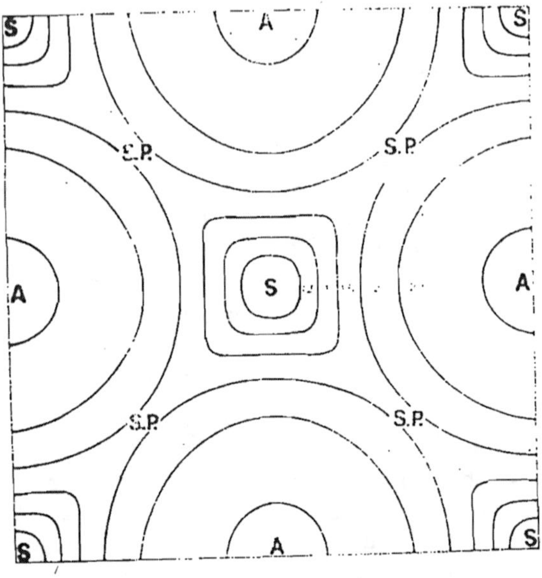

Fig. 16.2: Map of the adsorption potential energy surface for an He atom on the (100) face of solid Xe. Numbers on the energy contours give energies related to the ε_0 value for an isolated He-Xe pair of atoms.

A = surface atom, S = adsorption site S.P. = saddle point.

Ricca et al suggested that a better way to approach the concept of mobility, would be to use a quantum mechanical approach to calculate probability densities and vibrational energy levels for adsorbed gases by solving the Schrodinger equation for an atom in the already determined gas-solid potential field:

$$H\psi = -\frac{h^2}{2m}\nabla^2\psi + {}_mU_o(\bar{r})\psi = \varepsilon\psi \qquad (5)$$

where H is the Hamilton operator and ε and ψ are, respectively, the energy eigenvalue and eigenfunction of the adsorbed atom. On an ideal inert-gas crystal surface, ${}_{m}U_{o}$ (r) is a periodic function in two dimensions, and consequently ψ (r) and the probability density ψψ* (where ψ* is the complex conjugate of ψ), are also periodic. Thus calculating the local wave function for an adsorbed over a single site would be a useful first step, since the combination of such functions could give the wave function for the entire surface.

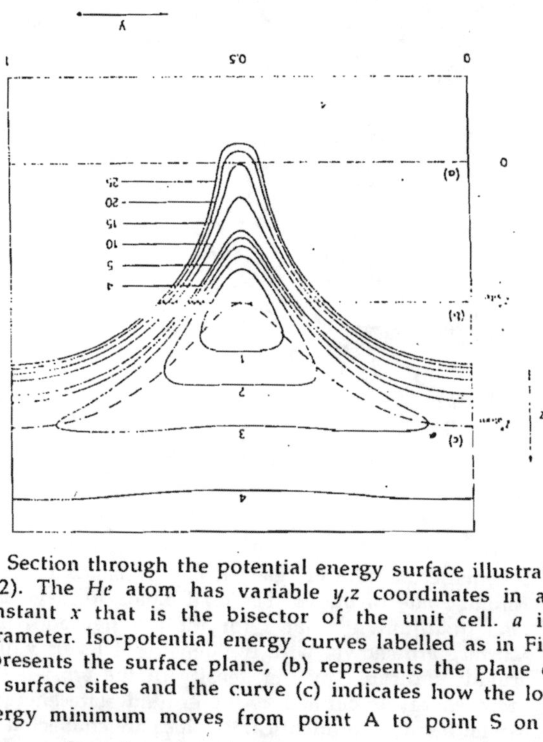

Fig. 16.3: Section through the potential energy surface illustrated in Fig. 16.2). The *He* atom has variable *y,z* coordinates in a plane of constant *x* that is the bisector of the unit cell. *a* is the cell parameter. Iso-potential energy curves labelled as in Fig. 4.26 (a) represents the surface plane, (b) represents the plane containing all surface sites and the curve (c) indicates how the locus of the energy minimum moves from point A to point S on Fig. 16.2.

The results of such calculations indicate that the energy levels for an atom on a surface containing N_s sites are concentrated into narrow bands containing *Ns* levels each. The width of a band depends upon the mass of the adsorbed atom as well as the parameters involved in determining ${}_{m}U_{o}$. Steel has given an energy band scheme based on Ricca's results, and this is shown in Fig. 16.4. This type of diagram would be valid for any adsorbed atom, and adsorbed helium is only a special case because the bandwidths of the low-lying states are much larger than for heavier atoms.

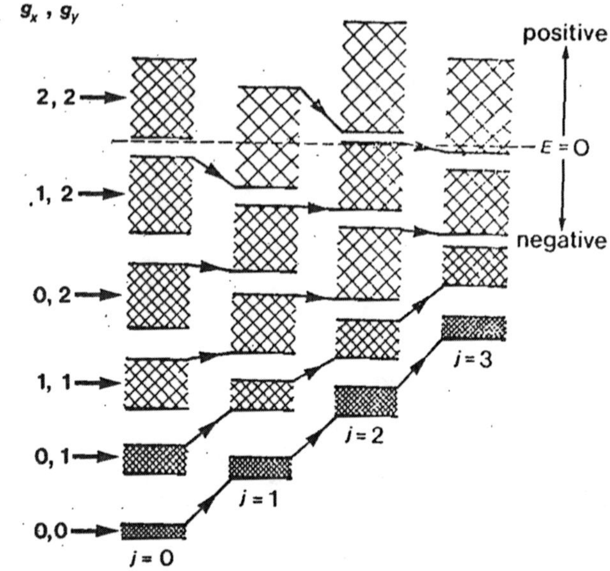

Fig. 16.4: Schematic diagram of selected energy bands for an adsorbed atom. Some of the translational bands for several values of the quantum number j for motion perpendicular to the surface are shown, with the density of levels within a band rughly indicated by the density of the cross-hatching. The zero energy is defined as the ground state energy of a free atom. Bands with identical g values are connected by arrows. (After Steele).

The results in term of probability density diagrams are shown in Fig. 16.5 for various inert gas atoms on the (111) face of a xenon crystal. These represent the fundamental 'bonding' state of two adjacent sites. A qualitative difference exists between the localised atoms of Ar and Ne and the mobile atoms of the He^4 and He^3. The 'orbitals' for the He atoms extend over both sites in contrast to the Ar and Ne cases.

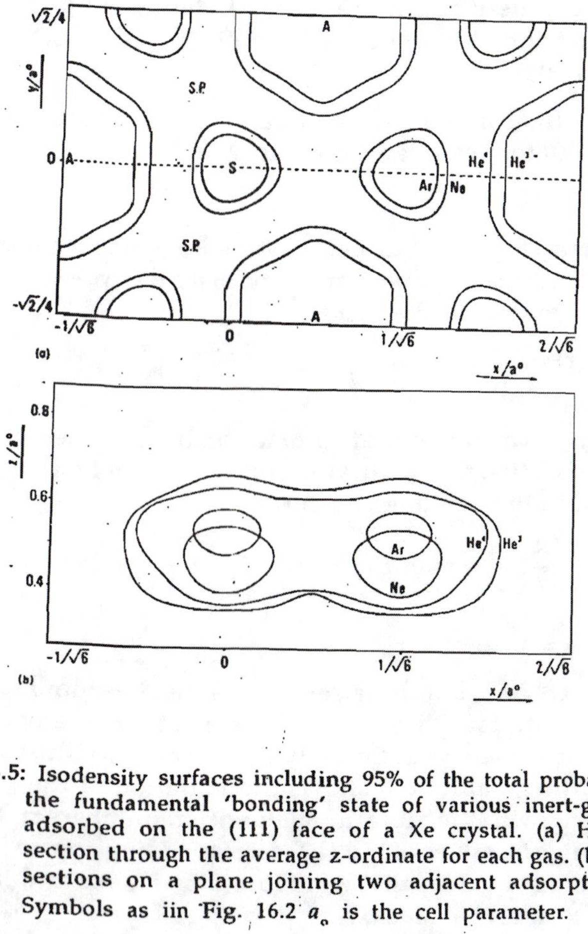

Fig. 16.5: Isodensity surfaces including 95% of the total probability for the fundamental 'bonding' state of various inert-gas atoms adsorbed on the (111) face of a Xe crystal. (a) Horizontal section through the average z-ordinate for each gas. (b) Vertical sections on a plane joining two adjacent adsorption sites. Symbols as iin Fig. 16.2 a_o is the cell parameter.

Adsorption potential maps; ionic solids

The calculation of adsorption potentials on an ionic solid ...esents considerable problems. The exact form of the equation ...o represent the dispersion force interactions between the ...g atom or molecule is somewhat difficult to decide, ... the added contribution to the total interaction of ...d due to the charges on the ions. This latter contribution is not an insignificant proportion of the total, and if realistic assessments are to be made, the complication of surface relaxation cannot be ignored. The data most suited to illustrating these points are those of House and Jaycock for the adsorption of inert gas atoms on ideal and relaxed alkali halide crystals.

The total interaction between an ionic crystal and an adsorbing atom can be written as

$$ {}_{m}U_{o} = U^{d}(r) + U^{e} \tag{6}$$

where $U^d(r)$ is the dispersion-repulsion interaction energy and can be estimated by the summation of the pairwise interactions over all the ions in the solid,

$$U^{d}(r) = \sum_{\text{all ions}} U(r)$$

although it was found that summation over the nearest 2000 ions was sufficiently accurate. U(r) was given by the '6-exp' equation in the Buckingham form

$$\frac{U(r)}{k} = \frac{\varepsilon/k}{1-(6/\alpha')}\left\{\frac{6}{\alpha'}\exp[\alpha'(1-(r/r_m))] - (r/r_m)^6\right\} \text{ for } r > r_{max} \tag{7}$$

$$\frac{U(r)}{k} = \infty \text{ for } r < r_{max} \tag{8}$$

system Ar/NaCl, but the latter case has not been found in any of these cases for which detailed calculations are available, and seems a somewhat unlikely occurrence. To illustrate the point further, Table 16.3 shows the variation of the fraction translating with the vibrational quantum numbers n_1 and n_2, over a range of values of $h\upsilon^{\perp}/kT$, corresponding approximately to the range of values that were found by House and Jaycock and Jaycock and Waldsax.

TABLE 16.3
Calculated values of the fraction of adsorbed atoms translating, R_t, for various values of $h\upsilon^{\perp}/kT$, and selected values of the quantum numbers n_1 and n_2. (Equations (4.48) and (4.49).

	$h\upsilon^{\perp}/kT$	1.00	0.75	0.50	0.30	0.10
n_1	n_2	R_t	R_t	R_t	R_t	R_t
1	1	0	0	0	0	0
	1.5	0.186	0.219	0.254	0.285	0.317
	2	0.269	0.321	0.378	0.426	0.475
	3	0.335	0.410	0.494	0.563	0.633
	5	0.364	0.460	0.571	0.666	0.758
	10	0.368	0.472	0.604	0.727	0.849
	100	0.368	0.472	0.607	0.741	0.905
2	2	0	0	0	0	0 0
	2.5	0.058	0.082	0.114	0.145	0.181
	3	0.090	0.132	0.186	0.240	0.301
	5	0.129	0.204	0.311	0.419	0.539
	10	0.135	0.223	0.364	0.525	0.713
	100	0.135	0.223	0.368	0.549	0.819
3	3	0	0	0	0	0
	3.5	0.020	0.036	0.060	0.087	0.122
	4	0.032	0.059	0.102	0.151	0.214
	5	0.043	0.084	0.154	0.236	0.341
	10	0.050	0.105	0.218	0.375	0.590
	100	0.050	0.105	0.223	0.407	0.741
5	5	0	0	0	0	0
	7.5	0.006	0.020	0.060	0.132	0.254
	10	0.007	0.023	0.076	0.182	0.378
	100	0.007	0.024	0.082	0.223	0.607

This table illustrates four important points:

(i) As n_1 increases the degree of localisation increases, that is, increasing translational barriers increases the degree of localisation;

(ii) For a fixed value of n_1, increasing values of n_2 increases the mobility of the adsorbed film, that is, increasing the width of the translational energy band increses mobility;

(iii) Decreasing the size of the vibrational quanta increases mobility;

(iv) Increasing temperature increases the film mobility.

On the basis of such calculations as are currently available, many cases would be expected to be pseudo-localised, and a few to correspond to the mobile and localised extremes. Of the cases so far discussed, H_2O on the surface would be localised (calculations suggest Rt~10-50), and Ar/NaC1 mobile, but of the other cases namely Kr/NaCl, Ar/KCl and Kr/KCl, all are pseudo-localised. It is interesting to note that inceasing the size of the adsorbed atom does not necessarily increase the degree of localisation. For example, Ar/NaCl is mobile and 40% of the adsorbed atoms translate for Kr/NaCl; but 29% of adsorbed Ar atoms on KCl translate, compared with 43% in the case of Kr/KCl.

17

Isosteric Enthalpy

Introduction

The Langmuir isotherm provides a simple model for adsorption—desorption equilibrium. However, there are several cases of adsorption for which this isotherm is inadequate. The chief fault with the Langmuir approach is that it assumes that all adsorption sites are equivalent. If that were the case, one would expect the enthalpy of adsorption to be independent of surface coverage. But this is not always true. Experiments show that the enthalpy of adsorption usually decreases with increased surface coverage. Such behaviour indicates that the sites on the surface are not energetically equivalent and that those with greater binding energy are occupied first. The enthalpy of adsorption at constant surface coverage is called the *isosteric* enthalpy.

Show that isosteric enthalpy is given by

$$\Delta H_{ads} = -RT^2 \left(\frac{\partial \ln P}{\partial T}\right) \tag{1}$$

The thermodynamic equilibrium is between molecules in the gas phase and on the solid surface. Hence from the chemical potential equation, we have

$$\frac{d\mu(\text{gas})}{T} = \frac{d\mu(\text{surface})}{T}$$

The chemical potential of the gas depends on pressure

and temperature, and the chemical potential of the surface molecules depends on temperature. Hence

$$R d \ln P - \frac{H(\text{gas})}{T^2} dT = - \frac{H(\text{surface})}{T^2} dT$$

from which we get

$$R\, d \ln P = \frac{[H(\text{surface}) - H(\text{gas})]}{T^2} dT = - \frac{\Delta H_{ads}}{T^2} dT$$

Rearrangement of this equation gives Eq. 1. This equation is quite similar to the Clausius-Clapeyron equation except for the sign of the enthalpy. The sign difference arises because the enthalpy in Eq. 1 refers to condensation and not to evaporation.

The Langmuir isotherm may be written in an alternative form:

$$\frac{\theta}{1 - \theta} = KP \tag{2}$$

which shows that, at constant θ, $d \ln K + d \ln P = 0$. Hence the isosteric enthalpy is given by

$$\left(\frac{\partial \ln K}{\partial T} \right) = \frac{\Delta H_{ads}}{RT^2} \tag{3}$$

This equation resembles the Gibbs-Helmholtz equation. Figure 17.1 gives examples of how enthalpy varies with surface coverage.

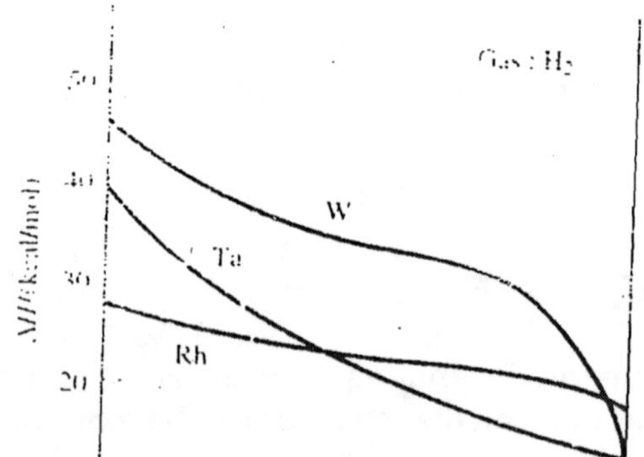

Fig. 17.1: Enthalpy of adsorption as a function of surface coverage.

18

Surface Energy of a Solid

1. Introduction

The newly formed surface of a liquid rapidly takes up an equilibrium conformation, whereas, as we have previously remarked, the same is not true of a solid surface. The latter is likely to have a considerable range of values of surface free energy, varying from region to region on the surface, and also at any one point, unlike the surface of a liquid, the surface tension need not be the same in all directions. Let us suppose that the surface tension can be resolved into two diretions at right angles, and that we can represent these partial surface tensions by γ^1 and γ^2. For an anisotropic solid, if the area is increased in two directions by dA_1 and dA_2, as shown in Fig. 18.1, then the total increase in available energy is given by the reversible work done against the stress γ^1 and γ^2. Thus

$$d(AA^S)_{T.V.n} = \gamma^1 \, dA_1 + \gamma^2 \, dA_2 \quad (1)$$

where A^S dénotes the available energy per unit area. If $\gamma^1 = \gamma^2 = \gamma$ then

$$\gamma = \frac{d(AA^S)_{T,V,n}}{dA} = A^S + A\left[\frac{dA^S}{dA}\right]_{T,V,n} \quad (2)$$

which can also be derived for an isotropic solid from equation (3)†

† The available energy for an open system may be written

$$dA = SdT - PdV + \gamma dA + \Sigma_i \mu_i \, dn_i \quad (3)$$

and at constant T, V and n this yields

$$\gamma = \left(\frac{\partial A}{\partial A}\right)_{T,V,n} \quad (4)$$

Then $dA = d(A^S A) = A^S dA + A dA^S$ (5)

which on substitution in equation (3) gives equation (2)

298.15 K ——and 77.5 K ——.

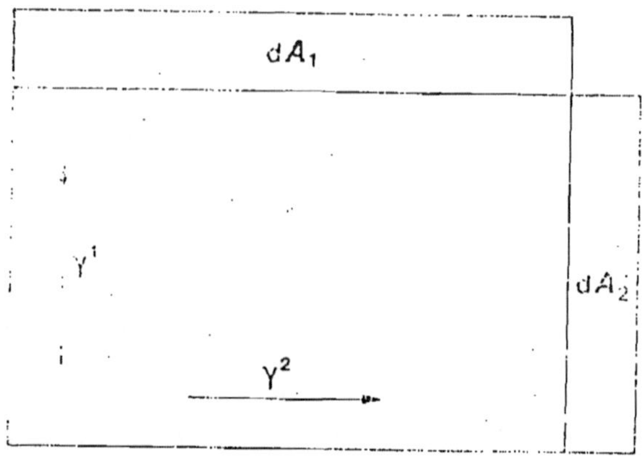

Fig. 18.1: Resolution of the surface tensions of an anisotropic solid in two directions.

In the case of a solid, however, $dA^S / DA = 0$ as a general rule, in fact it could only be so for an homotattic (ideal crystal-like) surface. Thus only if the surface achieved some uniform equilibrium state would $A^S = \gamma$, and in all other real cases A^s and γ will be different from their equilibrium values and different from each other. It cannot be legitimate therefore to call γ the specific surface free energy.

This problem can also be seen to have further implications. If we integrate equation (3) under conditions of constant T and V then

$$A = \gamma A + \Sigma_i \mu_i n_i \quad (6)$$

or

$$A^S = \gamma + \Sigma_i \mu_i \Gamma_i \quad (7)$$

Thus even if the conditions of an ideal surface could be maintained in making $dA^S/dA = 0$, then $\Sigma_i \mu_i \Gamma_i = 0$ must also be satisfied. The reader is recommended to consider this last point again after reading the section dealing with the calculation of surface energy.

This type of consideration of the surface in relation to Gibbs free energy raises certain problems. If one starts with the equation

$$dG = -SdT + VdP + \gamma dA + \Sigma_i \mu_i dn_i \tag{8}$$

then

$$\left[\frac{\partial G}{\partial A}\right]_{T,P,n_i} = \gamma \tag{9}$$

and therefore for an isotropic solid

$$\gamma = \frac{\partial(G^S A)}{\partial A} = G^S + A\left[\frac{\partial G^S}{\partial A}\right]_{T,P,ni} \tag{10}$$

which is analogous to equation (2). However, these concepts were introduced in this section using the ideas of reversible work. If, on the other hand, we start with

$$dG = -SdT + VdP - Ad\gamma + \Sigma_i \mu_i dn_i \tag{11}$$

then

$$\left[\frac{\partial G}{\partial \gamma}\right]_{T,P,n_i} = -A \tag{12}$$

which if we expand analogously gives

$$\left[\frac{\partial(G^S A)}{\partial A}\right]_{T,P,n_i} = -A\left[\frac{\partial y}{\partial A}\right]$$

or

$$G^S = -A\left[\frac{\partial \gamma}{\partial A} + \frac{\partial G^S}{\partial A}\right]_{T,P,n_i} \tag{13}$$

which is not similar in form to equation (2). On the other hand integrating equations 8 and 11 at constant T and P gives

$$G = \gamma A + \Sigma_i \mu_i n_i \tag{14}$$

$$G = -\gamma A + \Sigma_i \mu_i n_i$$

and therefore

$$G^S = \gamma + \Sigma_i \mu_i \Gamma_i$$

$$G^S = -\gamma + \Sigma_i \mu_i \Gamma_i \tag{15}$$

The concept of surface tension applied to a solid has important implications in relation to the Laplace equation 16. Let us consider for a moment a nearly spherical crystal such that we may write

$$\Delta P = 2\gamma / \mathrm{r} \tag{16}$$

The pressure differential ΔP will result in compression of the crystal. Thus if β is the compressibility, the we may write the approximate relation

$$\frac{\Delta V}{V} = 3\frac{\Delta r}{r} = -\Delta P\beta \tag{17}$$

or

$$\Delta \mathrm{r} = -2\beta\gamma/3 \tag{18}$$

This effect has been investigated for magnesium (II) oxide, using X-ray powder diffraction measurements. Values of Δr near the expected value of 0.06 nm, corresponding to an 0.1% change in lattice distance, were found for a 60 nm radius crystallite using a calculated value for γ of 6.573 N m^{-1}. Reinvestigation of the situation by Guilliatt and Brett indicates that any adsorbed material such as water vapour produces lattice dilation, whereas in the absence of an adsorbed film the expected lattice contraction is observed, the relative magnitude of which increases with decreasing crystallite size, and is substantially in agreement with the theoretical predictions of Anderson and Scholtz. It is worth noting that surface dilation would imply a negative value for γ in euation(18).

Shuttleworth set out to show that equating of surface tension and surface stress had very real conceptual problems for crystal line solids.* He defined surface stress analogously to stress in bulk elasticity. If one imagines a cut is made perpendicular to the surface of the crystal and extending only

* Gibbs has pointed out that the surface tension, or precisely the surface energy, is the work necessary to form the surface, whereas the surface stress represents the work necessary to stretch the surface.

a little way into it, then in order for the surface to achieve an equilibrium configuration, and no additional stresses to appear in the bulk of the crystal, then a surface stress must be present, and could be achieved by local relaxation of the crystal structure. Since this surface stress will have at least the symmetry of the crystal face, then the surface tension can be taken to specify it when the surface has a three-fold (or greater) axis of rotational symmetry. In this case the normal stress components across all lines in the face are equal and all shear stresses are zero, so that the normal stress components are equal to the surface tension.

In a face-centred cubic crystal the (111), (100) and (110) faces have six-fold, four fold and two-fold axis of symmetry respectively. The shear component of the surface stress will be zero for the (111), (100) faces, but not for the (110) face. Shuttleworth attempted calculations for (100) faces of inert gas crystals and alkali halide crystals at oK, expressing the surface energy as

$$U = U' + U'' \tag{19}$$

where U' is the value derived from the surface energy of the crystal before any rearrangement of the surface atoms relative to ideal positions occurs, and U'' that due to the surface energy decrease on relaxation. He concluded that the surface stress was —130 mN m^{-1}. On the other hand Gibbs has shown that the surface tension must be positive as a condition for stability, and Dunning quotes a value when reviewing the work of Shuttleworth and Lennard-Jones and Dent, of 155 mN m^{-1}.

The mechanism by which surface stress could be relieved has been discussed by Herring and shown diagrammatically by Dunning. Suppose a cube of ideal crystal is subjected to a compressive surface stress, then its deformation (Fig. 18.2(a)) is equivalent to applying traction to each edge of the cube. This stress could be relieved by the presence of rows of dislocation just below the surface (Fig. 18. 2(b)), since dislocations would have the effect of stretching a surface without increasing its area, or by vacancies actually in the surface (Fig. 18.2 (c)). Such processes must involve increases in surface energy and the generation of surface heterogeneity.

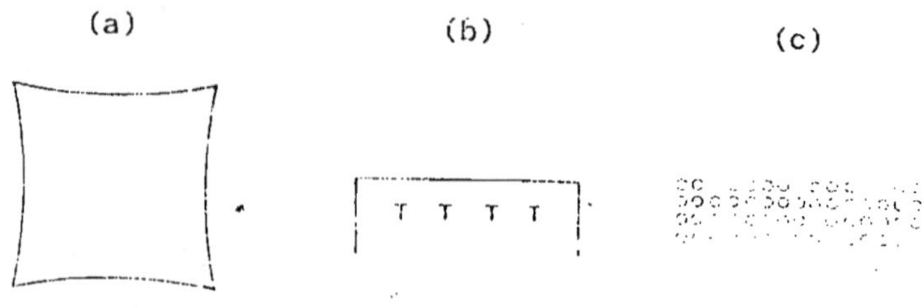

Fig. 18.2: Herring's mechanism for the relief of surface stress by dislocations.

2. Calculated surface energy values

Because of the impossibility of measuring surface tension values for solid surfaces, except in a few abnormal circumstances calculated surface energy values for solids take on a significance that they lack in the case of liquid interfaces. We have already referred to the early work in this field of particularly Lennard-Jones and Dent, but they, and the majority of other workers before about 1960, were considerably limited in terms of what they could attempt, by severe computational difficulties. The advent of the modern computer changed this situation, and we can illustrate the impact that these made by looking first at the (100) face of sodium chloride crystals.

The Lennard-Jones and Dent model, although extremely elegantly treated, was restricted by the face that they considered an ideal crystal, in which the ionic positions at the surface were identical to those achieved in the bulk crystal. This is obviously extremely improbable, and the first attempt to understand relaxation processes was due to Verwey. He assumed a bulk lattice spacing of 0.281 nm, and concluded that in the outermost layer the chloride ions moved out from the predicted ideal plane position to 0.286 nm from the plane of the ideal bulk crystal immediately beneath, and that the sodium ions moved inwards to 0.266 nm from the same plane, as illustrated in Fig. 18.3. Thus the plane of the outer chlorine ions was 0.020 nm farther out than the plane of the outer sodium ions, giving rise to a surface double layer. This differential effect is the result of the fact that the larger negative ions are more polarisable than the smaller positive ions, and that consequently there will be a larger induced electric dipole moment in the chloride ion than in the sodium ion. Thus the outer chloride ions will move so as to increase the distance of the positive end of the dipole from the plane of the sodium ions and decrease the distance of separation of the negative end.

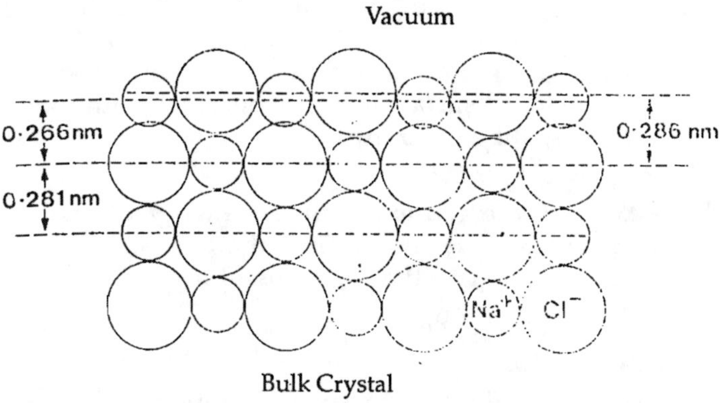

Fig. 18.3: Verwey's single layer relaxation model of sodium chloride. In his calculations, compared with a bulk crystal layer separtion of 0.281 nm, the outer sodium ions moved inwards by 0.015 nm and the outer chloride ions outwards by 0.005 nm. The circles represent approximate ionic radii.

There would seem no logical reason for supposing that only the outer plane should relax, and that some larger number of layers will in fact show significant displacements. Benson *et al*, as part of a general study of relaxation in the surfaces of alkali halide crystals, showed the outer five layers were all significantly displaced from the ideal crystal position, and that only from the sixth layer in from the surface onwards was it fair to consider the crystal as bulk ideal crystal.

The method of calculation was to use numerical methods to minimise the total energy of the system by allowing ions in the outer five layers to move to new equilibrium positions. The interaction energy of two ions, *i* and *j*, was taken to be

$$u_{ij} = u^0_{ij} - e_i(r_{ij}.\mu_j)r_{ij}^{-3} + e_j(r_{ij}.\mu_i)r_{ij}^{-3} - 3(r_{ij}.\mu_i)(r_{ij}.\mu_{ij})r_{ij}^{-5} + (\mu_i.\mu_j)r_{ij}^{-7} \quad (20)$$

where *e* and *u* denote charge and dipole moment respectively and r_{ij} is the position vector of the ion i relative to the ion *j*, and

$$u^o_{ij} = e_ie_jr_{ij}^{-1} - c_{ij}r_{ij}^{-6} - d_{ij}r_{ij}^{-8} + b_{ij}\,exp(-r_{ij}/\rho) \quad (21)$$

is the Born-Meyer form of the potential energy function and

represents the sum of coulombic, van der Waals and repulsive contributions. The other terms in equation (20) are the energies of the charge-dipole and dipole-dipole interactions arising out of the polarisation of the ions, and the exponential term is the repulsive energy involving the 'hardness' parameter, ρ. Since the area occupied by an ion pair in the surface is $2a^2$, where a is the nearest neighbour separation in the bulk crystal, then the correction to the surface energy (compared with that of the ideal crystal) willbe given by

$$\Delta\sigma_{(100)} = \Delta U_{min}/2a^2 \tag{22}$$

where ΔU_{min} is the reduction in the energy of the crystal brought about by relaxation. Thus the true surface energy, $\sigma_{(100)}$, will be related to that of an ideal crystal, $\sigma^0_{(100)}$, by

$$\sigma_{(100)} = \sigma^0_{(100)} + \Delta\sigma_{(100)} \tag{23}$$

$$= 210.9 - 107.4 = 103.5 \text{ mJ m}^2$$

assuming the outer five layers relax. The significance of increasing the number of layers allowed to relax inthe calculation can be seen in Fig. 18.4.

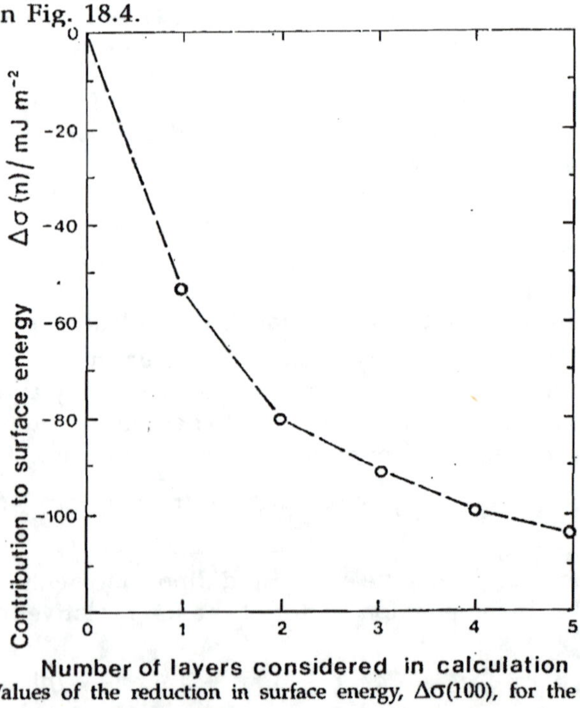

Fig. 18.4: Values of the reduction in surface energy, $\Delta\sigma(100)$, for the (100) face of NaCl, as a function of the number of layers assumed to relax.

Benson *et al* data on the positional changes of the ions was represented by them as shown in Fig. 18.5 for the case where the outer five layers are allowed to relax. The most significant factor is that whilst the chloride ions are increasingly displaced away from the surface as the surface is approached from the bulk of the crystal, the displacement of the smaller sodium ions alternates. However, the degree of disorder or distortion this relaxation process actually produces in the surface can be more readily seen in Fig. 18.6, in which the circles represent approximate values for the bulk crystalline ionic radii. It can be readily appreciated from the size of

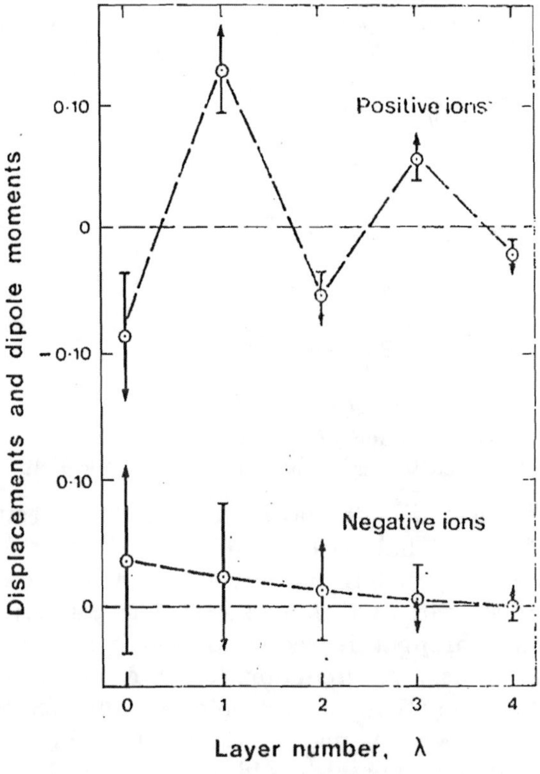

Fig. 18.5: Equilibrium configuration of the first five layers, for the (100) face of NaCl. Displacements in units of *a*. Positive values indicate normal displacements outwards. The direction and magnitude of the induced dipole moments (Debye units) are indicated by the arrows, but for the negative ions they are scaled down by a factor of ten.

$\Delta\sigma_{100}$ and the positional shifts of Fig. 18.6, that crystal surfaces must generally be very different from bulk crystal planes, and this would be expected to produce considerable changes in the adsorption behaviour from that predicted on the basis of ideal crystal geometry.

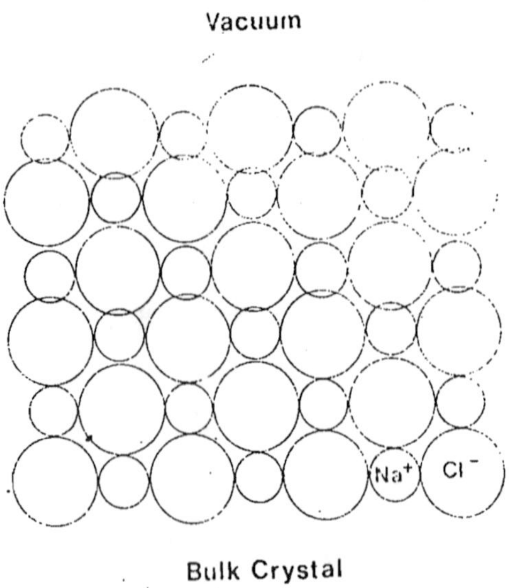

Fig. 18.6: Surface relaxation of the (100) face of sodium chloride crystals. The circles represent approximate ionic radii.

Theoretical studies of relaxation at most crystal surfaces other than the alkali halides are still lacking. There is no real reason why any completely ionic crystals could not be treated in a similar way to that which Benson *et al* have adopted, but for crystals exhibiting a degree of covalent character, silver iodide for example, a different approach is needed. The most probable route to solving such problems would seem to lie through the use of wave-mechanical methods, but it is not clear at the moment how this could be done, and the authors are not aware of any successful attempt. Crystals of inert gases held together by van der Waals forces alone, can of course be dealt with in an analogous manner to that of Benson *et al.*

19

Heat of Wetting

Introduction

Heat is evolved when a solid is immersed in a liquid, and this is the *integral* heat of wetting (or immersion). In the experimental technique for measuring this heat the solid, usually in powder form, is kept in a vacuum and broken into the liquid under its own vapour. The quantity of heat evolved, besides being a function of the nature of the solid and liquid and the intermolecular forces involved in creating the new interface, is dependent on the purity of the liquid and the pretreatment of the solid, and without appropriate attention to these factors the measured heat could be quite misleading. For example, the heat of wetting of titanium (IV) oxide (anatase) in dry benzene is 150mJ m^{-2} and in water 520 mj m^{-2}. Addition of 180 ppm water to benzene raises the heat to 450 mj m^{-2}, the water having a much higher affinity for the oxide surface than for the benzene. These figures relate to a powder from which surface adsorbed water has been removed, and for titanium (IV) oxide this requires heat treatment in vacuum at ≥ 150°C. A sample evacuated at room temperature and immersed in dry benzene gives a heat of wetting of about 200 mj m^{-2}, which of course is not the true value for the titanium (IV) oxide-benene system. There is an abundance of heat of wetting data in the literature for a variety of solid-liquid pairs, but although for each experiment precision is high there are significant variations from one laboratory to another—just as in the case of contact angle measurement the 'microscopic' state of the system is all important, and solid-liquid interfacial phenomena are critically dependent on impurities, pretreatment of solid etc.

The enthalpy change (per unit area of solid) in the immersional process is given by

$$\Delta_w H = h^S_{SL} - h^S_S - n^S_L h_L \qquad (1)$$

where h^S_{SL} and h^S_S are respectively the enthalpies of the solid-liquid and solid-vacuum interfaces, h_L is the enthalpy per mole of liquid, and n^s_L is the number of moles of liquid adsorbed at the interface, that is, the a different state from those in bulk liquid. The term $n^S_L h_L$ accounts for the loss of enthalpy in the bulk liquid due to molecules being transferred to the interface on adsorption.

The enthalpy of the solid-liquid interface h^s_{SL} is made up of two terms:

(a) the enthalpy h^s_a of the adsorbed layer of liquid molecules when in equilibrium with saturated vapour of the liquid, and

(b) the enthalpy h^s_L of formation of the interface between the adsorbed layer and bulk liquid, and this is taken as equal to the enthalpy of the liquid-vapour interface, that is, the heat evolved on destruction of unit area of liquid surface, as given by

$$h_L = \gamma^{LV} - T(\partial\gamma^{LV}/\partial T). \qquad (2)$$

which can be derived for a closed system at constant P and T. Strictly this assumption means that the adsorbed layer at the interface must be sufficiently thick that the energies of the surface of the layer are not affected by the the underlying solid, that is, it is a duplex film. Molecules adsorbed from the vapour phase on such a film do so with a heat evolved equal to the heat of liquefaction of the liquid. For a number of systems this occurs at about monolayer coverage of the solid surface by liquid molecules, but how much adsorbate is required to mask the effects of the solid depends on the nature of both adsorbate and adsorbent.

There is still considerable debate on the range of the force associated with the solid surface, and the number of layers of adsorbed molecules affected. For a flat surface, oxidised aluminium foil, contact angle and adsorption data demonstrate that a monolayer (in vertical orientation) of a straight-chain alcohol containing three or more carbon atoms is sufficient to mask the high energy character of the solid surface. The extensive work of Derjaguin has shown that the action of surface forces extends over large distances in polar liquids, and the curious phenomena associated with water in fine glass capillaries have been related to the long-range effect of the solid. Differential heats of adsorption (from vapour adsorption experiments) usually decrease steadily with increasing amount adsorbed, and where physical adsorption is involved tend to the heat of liquefaction of the adsorbate as the pressure approaches staturation. So the use of equation (2) requires consideration of the characteristics of the system under study e.g. one might hesitate for the case of an oxide in water but have more confidence with a hydrocarbon and a pure carbon block.

Hence we write

$$h^{S}_{SL} = h^{S}_{a} + [\gamma^{LV} - T(\partial\gamma^{LV}/\partial T)] \quad (3)$$

and from equation (1) and (3)

$$\Delta_w H = h^{S}_{a} + [\gamma^{LV} - T(\partial\gamma^{LV}/\partial T)] - h^{S}_{S} - n^{S}_{L} h_{L} \quad (4)$$

Now consider the immersion of a solid that has been equilibrated with the vapour of the liquid so that the surface now contains an adsorbed layer of an extent that depends on the pressure of the liquid to which it was exposed. If the adsorbed film is of the duplex type, that is, there is no furher adsorption of liquid molecules on immersion, the difference between this enthalpy change $\Delta_w H_D$ and that for an outgassed surface is a direct measure of the difference between the enthalpies of molecules in the adsorbed and bulk liquid states, hence of the enthalpy of adsorption $\Delta_a H$. Thus

$$\Delta_a H = \Delta_w H - \Delta_w H_D \quad (5)$$

and

$$\Delta_a H = h^{S}_{a} + [\gamma^{LV} - T(\partial\gamma^{LV}/\partial T)] - h^{S}_{S} - n^{S}_{L} h_{L} - \Delta_w H_D \quad (6)$$

It is often assumed that $\Delta_w H_D$ may be replaced by h_L, but once again it is necessary for the adsorbed film to be sufficiently thick for it to be liquid-like in terms of its enthalpy. We may then write

$$\Delta_a H = \Delta_w H - [\gamma^{LV} - T(\partial\gamma^{LV}/\partial T)] \quad (7)$$

so that adsorption enthalpies are obtained from heat of immersion data combined with appropriate surface tension data, although $\Delta_w H_D$ is readily measured. Immersional heats as a function of pre-coverage are of particular interest since such data provide useful information on the heterogeneous nature of the solid surface amount.

The heat of immersion technique also provides a direct method of evaluating ${}^d\gamma^S$, the dispersion contribution to the surface free energy of a solid. For the immersional process we may define the free energy change (per unit area of solid) as

$$\Delta G^{SL} = \gamma^{SL} - \gamma^{SV} \quad (8)$$

and

$$\Delta G^{SL} = \gamma^{LV} - 2\sqrt{({}^d\gamma^L\,{}^d\gamma^S)} \quad (9)$$

Since the enthalpy and free energy changes are related by

$$\Delta_w H = \Delta G^{SL} - T(\partial(\Delta G^{SL})/\partial T) \quad (10)$$

then

$$\Delta_w H = \gamma^{LV} - 2\sqrt{({}^d\gamma^L\,{}^d\gamma^S)} - T\left[\frac{\partial({}^d\gamma^L)}{\partial T} - 2\sqrt{({}^d\gamma^L)}\left(\frac{\partial\sqrt{({}^d\gamma^S)}}{\partial T}\right) - 2\sqrt{({}^d\gamma^S)}\left(\frac{\partial\sqrt{({}^d\gamma^L)}}{\partial T}\right)\right] \quad (11)$$

and values of ${}^d\gamma^S$ may be calculated from equation (11) — data for the liquid γ terms are well established, but little is known about the temperature coefficient of the dispersion contribution which nevertheless can be estimated from the fourth power of the density without introducing too much erro. Zettlemoyer used this method to calculate ${}^d\gamma^S$ from the interaction between the graphitised carbon black Graphon and some short chain hydrocarbons for which system only dispersion forces would be expected to operate. His data (for 25°C), given in Table 19.1, average to a value very close to that (70 mJ m^{-2}) established by other techniques.

TABLE 19.1
Calculation of $^{d}\gamma^{S}$ for Graphon from heat of wetting data

	$\Delta_w H/mJ\ m^{-2}$	$^{d}\gamma^{S}/mJ\ m^{-2}$
hexane	103	65
heptane	112	68
octane	127	78.5

The experimental heat of wetting data in Table 19.1 show an increase with chain length of the straight chain alkane; data are available for the series and show a two-fold increase in heat per unit surface area from hexane to heaxadecane. Furthermore, for hexadecane the heat of wetting decreases with increasing temperature, and to explain these observations it has been postulated that close to the freezing point of the liquid the layer of liquid adjacent to the solid surface has properties intermediate between those of the liquid and crystalline hydrocarbon. This concept is supported by measurements of the volume changes that accompany the wetting of Graphon by straight chain alkenes, and demonstrate a sharp increase in density parallel to a similar effect in heat of wetting as the temperature approaches the freezing point of the hydrocarbon. It would appear that surface forces induces some structural order into the layers of hydrocarbon close to the interface with the long axes of the molecules lying parallel to one another, as in a crystal — a 'pre-freezing' phenomenon. How far this extends into the bulk liquid is not known, and would depend on the chemical nature of the solid/liquid pair, but in the Graphon/alkane case it seems to be a significant distance sufficient to make possible experimental measurements of parameters arising from the effect. A similar phenomenon occurs with the primary aliphatic alcohols but not with the pure monocarboxylic acids, and the reason for this difference in behaviour is not completely understood.

It is worth mentioning at this stage that there has been a great deal of controversy over whether a solid can have a long-range influence on the structure of liquid in contact with it, particularly with respect to water. Numerous experiments suggest that water in proximity to a solid surface is different from the bulk. The first layer adjacent to the surface is subject to restricted molecular motion, and the effect can appear to be transmitted over many molecular diameters via hydrogen bonding. The layer of water so affected has been called 'vicinal water', and the experiments which support or otherwise the presence of such a water layer have recently been reviewed by Peschel and Belouschek. There is no unambiguous information on the thickness of the layer — it may extend up to 10 nm. But if such water exists it will be of vital importance to the activity of biological cells. Cells and tissues are abundant in interfaces, and these must exert a marked influences on adjacent water structure.

We have already seen that the forces involved in the interaction between liquid and solid are of several kinds (dispersion, polar, etc.), and for any particular solid the heat of immersion might be expected to depend on the chemical nature of the wetting liquid. For Graphon, however, the interaction is principally through dispersion forces and the heat of wetting for a series of polar and non-polar molecules is contant within experimental error (Table 19.2). In contrast the heats for rutile titanium (IV) oxide are markedly dependent on the nature of the liquid, and the polarity of the latter is obviously important, and such data may be used to evaluate the field strength of the solid surface.

The net energy of adsorption of n^s_L mol m^{-2} is given, if the volume changes are negligible, by

$$\Delta_w H - \Delta_w H_D \cong n^S_L (U_a - U_l) \tag{12}$$

where subscripts *a* and *l* refer to adsorbed and liquid states. It is assumed that interactions between molecules in the adsorbed layer are the same as those in bulk liquid, so that only the surface/adsorbed molecule interaction is expressed in equation (12). We may divide the total interaction energy into three parts, being the contributions to London dispersion forces (U_d), to the polarisation of adsorbed molecules by the surface ($U\alpha$), and to the interaction between the permanent dipole in the adsorbate and the electrostatic field of the surfae ($U\mu$).

$$U_a - U_l = U_d + U_\alpha + U_\mu. \tag{13}$$

Now

$$U\mu = F\mu \tag{14}$$

where F is the average electrostatic field strength of the surface, and μ the dipole moment of the adsorbate. Hence, combining equations (12), (13), and (14)

$$\Delta_w H - \Delta_w H_D = n^S_L (U_d + U_\alpha) - n^S_L F\mu \tag{15}$$

and a plot of the left-hand side versus μ should be linear with a slope from which the field strenth may be calculated. For a series of organic molecules only differing in the polar group, for example the butyl derivative given in Table (19.2), it may be assumed that for a monolayer covered surface $\Delta_w H_D$ is approximately constant since this surface for all the derivatives is essentially paraffinic in nature; hence a plot of $\Delta_w H$ versus μ should give the same result. Chessick *et al* have published such a plot for rutile titanium (IV) oxide using experimental heat of immersion data (Table 19.2) and these are illustrated in Fig. 19.1. From the slope an an effective field strength of 820 MV m^{-1} was calculted for the rutile surface. For Graphon and Teflon there is an undetectable change in net interaction energy with dipole moment from which it may be assumed that the electrostatic field strength of these non-polar surfaces is zero.

TABLE 19.2

Heats of immersion of Graphon and rutile titanium (IV) oxide ($\Delta_w H$/mJ m^{-2}) at 25°C

	Rutile	*Graphon*
methanol	426	110
ethanol	397	110
1-butanol	410	114
2-butanol	415	115
butanal	556	-
1-nitropropane	664	-
1-aminobutane	330	106
1-chlorobutane	502	106
butanoic acid	506	115
hexane	135	103
octane	140	127

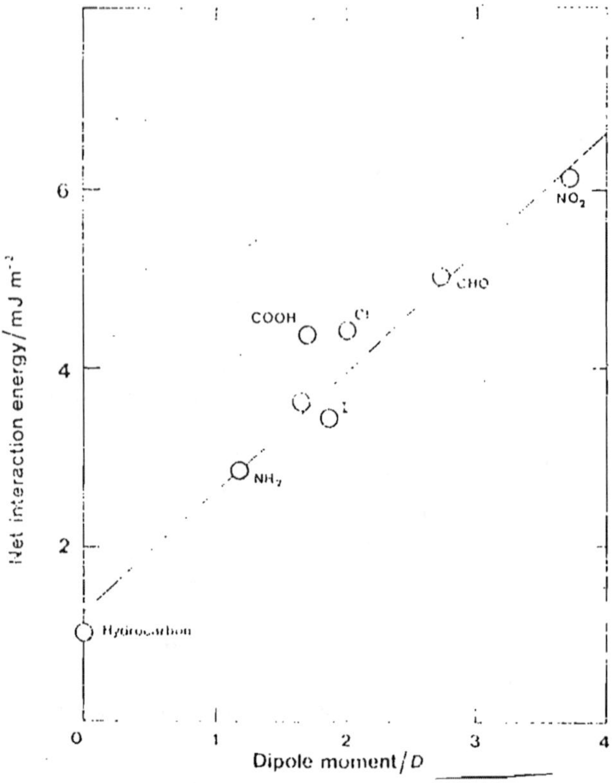

Fig. 19.1: Heat of immersion of rutile titanium (IV) oxide in various liquids as a function of the dipole moment of the liquid phase (from Chessick *et al*).

From F it is possible to calculate U_μ using equation (14) and U_α using

$$U_\alpha = -F^2{}_\alpha/2 \qquad (16)$$

where α is the polarisability of the adsorbate, giving U_d by difference. Data for rutile and Graphon are given in Table 19.3.

In a later publication Zettlemoyer estimated, using Fowkes' approach, that of the 94 mJ m^{-2} calculated for U_d for 1-butanol on rutile 25 mJ m^{-2} are due to hydrogen-bonding.

The heat of immersion measurement is a potential source of fundamental data, and although high-precision calorimeters are now available commercially, the extent to which they have been used in assessing basic interfacial interctions has as yet been relatively small. Furthermore, it normally involves a high area powder, and unfortunately in most cases it is not possible to make accurate measurements of contact angle for such material. So heat of immersion and contact angle measurements are made on differen materials, which may well have the same chemical composition but are subject to all the uncertainties mentioned earlier in terms of surface preparation.

Table 19.3

Interaction energies/mJ m^{-2} for Graphon and rutile titanium (IV) oxide.

	U_d	U_α	U_μ
hexane on rutile	62	30	0
hexane on Graphon	67	0	0
1-butanol on rutile	94	21	246
1-butanol on Graphon	66	0	0

20

Thermodynamic Data from Heats of Immersion

Introduction

The analysis of the various enthalpy terms involved when a solid is immersed in a pure liquid is readily extended to solutions by taking account of the fact that the various components of the solution may be present in the adsorbed layer, and that on adsorption a concentration change takes place in solution.

Consider a two-component solution containing initially n_1 and n_2 moles of components 1 and 2 respectively, and after immersion there are n^S_1 and n^S_2 moles of these components in unit area of the adsorbed layer. By simple addition of the enthalpies of the two solution components and of the outgassed solid before and after immersion, the total change in enthalpy $D_w H_s$ during the immersional process is given by

$$\Delta_w H_S = h^S_{S1} - h^S_S - n^S_1 \bar{h}_1^{1} - n^S_2 \bar{h}_2^{1} - n_1(\bar{h}_1^{1*} - \bar{h}_1^{1} + n_2(\bar{h}_2^{1*} - \bar{h}_2^{1}) \quad (1)$$

where h^S_{S1} is the enthalpy of the *solid-solution* interface, and $\bar{h}_1^{1*}$ the partial molar enthalpy of molecules 1 in the bulk solution before adsorption and $\bar{h}_1^{1*}$ after adsorption. The terms within the square brackets represent the enthalpy change in bulk solution due to the change in concentration.

For a dilute solution the terms in the square brackets may be neglected, and the heat of immersion is then given by

$$\Delta_w H_S = h^S{}_{S1} - h^S{}_S - n^S{}_1 h_1{}^{-1} - n^S{}_2 h^{-1}{}_2 \quad (2)$$

and the differnetial heat of adsorption of component 1 is

$$\Delta \overline{H}_1 = \overline{h}_1^S - \overline{h}_1^{-1} = \left(\frac{\partial \Delta_W H_S}{\partial n^S{}_1}\right)_{P,\,T,\,n^S{}_2} \quad (3)$$

which can be derived from immersional heats and an adsorption isotherm.

An alternative approach considers that as for pure liquids, $h^s{}_{S1}$ is made up of two parts, that of the adsorbed layer $h^s{}_a$ and that for interface formation $h^S{}_L$. It is assumed that the molecular interactions in bulk solution are the same as those in the adsorbed layer, and need not be included in the total enthalpy change. Taking the fraction of surface covered by component 1 and θ then the enthalpy change for this component is (a) for adsorption $\theta(h^s{}_{SI} - h^s{}_s)$ and (b) for interface formation $\theta(h^S{}_{11} - h^S{}_{SI})$, where $h^S{}_{SI}$ is the enthalpy of surface covered with molecules of component 1, and $H^S{}_{11}$ is the enthalpy of the adsorbed layer (of 1) in contact with solution. Similarly for component 2 we have $(1-\theta)\,(h^S{}_{S2} - h^S{}_S)$ and $(1-\theta)\,(h^S{}_{21} - h^S{}_{S2})$. The total change on immersion is given by

$$\Delta_W H_S = \theta(h^S{}_{S1} - h^S{}_S) + (1-\theta)\,(h^S{}_{S2} - h^S{}_S) + \theta\,(h^S{}_{IS} - h^s{}_{SI}) + (1-\theta)(h^S{}_{2S} - h^S{}_{S2}) \quad (4)$$

Equation (4) is analogous to (1) except the solution terms have been excluded. Young, Chessick and Healey verified the validity of equation (4) using the Graphon 1-butanol/water system for which adsorption of 1-butanol reaches saturation at low concentration. Data for the individual terms on the right hand side were obtained from adsorption experiments and from the immersional heats for the individual components, and allowance was made for the enthalpy of dilution which they determined from heat of solution experiments. The agreement between the measured heats of immersion for the system and those calculated using equation (4) is excellent, and suggests that there is no appreciable contribution to the heat of immersion from interaction between molecules in the adsorbed layer, that is, the adsorption can be treated as the sum of independent effects.

When more concentrated solutions are considered, and particularly for the whole mole fraction range of two miscible liquids, any assumptions that lead to the elimination of solution enthalpy terms would, other than in exceptional cases, appear to be invalid. Everett has analysed the case for a perfect liquid mixture, and assuming monolayer adsorption shows that the heat of immersion in the mixture is related to those for the individual components by

$$\Delta_W H_S = x^S_1\, \Delta_W H_1 + X^S_2\, \Delta_W H_2 \qquad (5)$$

where x_1^S and x_2^S are the mole fractions in the adsorbed layer. Data on the charcol/benzene/cyclohexane system show good agreement with equation (5) over the entire concentration range. Extra terms must be included for solutions not exhibiting perfect behaviour but such terms are difficult to evaluate.

Furthermore the validity of assuming a monolayer is subject to debate as been said earlier, and some more recent heat of immersion data on the apparently simple system Graphon/heptane/hexadecane strongly suggest that more than one layer of adsorbed molecules are being affected by the underlying surface. Heats of immersion were measured over the whole mole fraction range at temperatures between 10° and 45°C, and are shown in Fig. 20.1. From a comparision of these results with adsorption from soution data it was concluded that the plateaux correspond to the completion of a monolayer of hexadecane molecules oriented parallel to the solid surface. The exeess heats at higher solution concentrations are attributed to the formation of a more structured phase in the adjacent surface layers, since as the temperature is lowered the excess heat increases at any one solution concentration. This is the same concept as discussed earlier as a 'pre-freezing phenomenon observed with straight chain alkanes and Graphon. The effect shows up clearly in the heat of immersion experiment, but is not recognised from adsorption data, indicating that in the surface layers affected the solution concentration is very close to that of the bulk liquid phase.

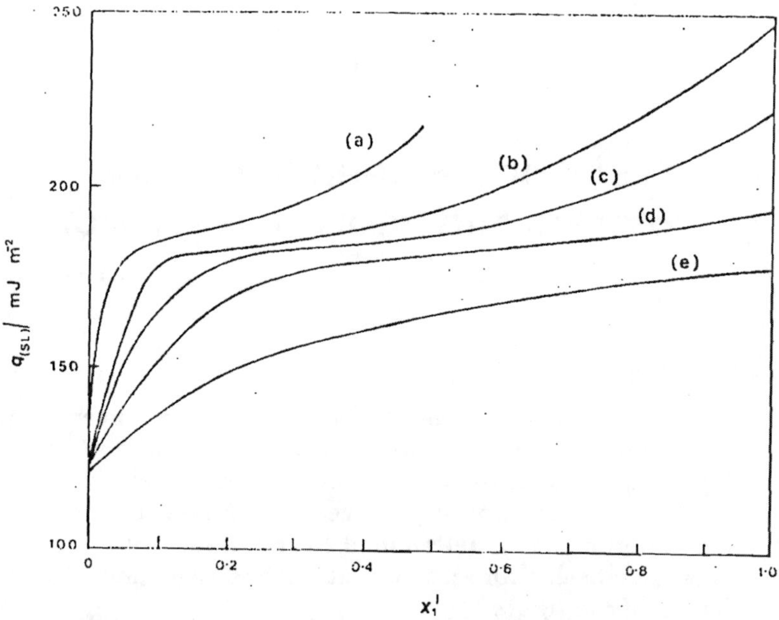

Fig. 20.1: Heats of immersion for Graphon in heptance/hexadecane solutions at 10°C (a), 20°C (b), 25°C (c), 35°C (d) and 45°C.

21

Relation between Adsorption Potential and Heats of Adsorption

Introduction

The potential energy curve for adsorbate-adsorbent interaction, as illustrated in Fig. 21.1 is the result of the sum of short-range Born repulsive forces and longer range van der Waals attractive forces, but the relationship between potential energy curves and experimentally determined heats of adsorption is not usually that clear, for reasons that can be best explained with the aid of the figure.

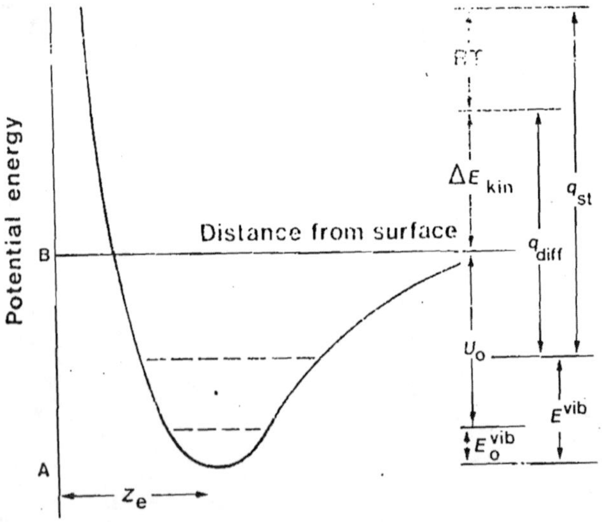

Fig. 21.1: The potential energy of interaction of an adsorbate molecule with the adsorbent, and the relationship to the heats of adsorption.

Firstly there are two ways of fixing the zero of the potential energy scale. The minimum value in the potential energy well, level 'A' in the figure, may be taken as zero, or the level 'B', that is, the potential energy when the adsorbate molecule is a large distance away from the surface, may be chosen as zero: the latter is a common choice when potential energy calculations are made. The relationships between the various quantities remain unaltered by this choice.

It is not possible for a molecule to remain stationary at the distance z_e, at the bottom of the potential energy well. Even at 0 K it will possess vibrational energy, and E_0^{vib} is the vibrational energy, at this temperature. When the temperature is increased it is possible to excite it to a higher vibrational level, and at the temperature T, this is shown in the figure as E^{vib}. Thus the problem in relating the differential heat of adsorption, q_{diff}, and the isosteric heat of adsorption, q_{st}, to the adsorption potential, U_o, is that usully the vibrational frequencies (and therefore E_0^{vib} and E^{vib}) are unknown.

The isosteric heat of adsorption is determined from a family of adsorption isothems at various temperatures, since

$$(\partial(\ln P)/\partial T)_{na} = (\partial(\ln P)/\partial T)_{\theta} = q_{st}/RT^2 \qquad (1)$$

where P is the equilibrium gas phase pressure, and n_a is the amount adsorbed. If the temperature range is small, then constant n_a is equivalent to constant fractional coverage, θ. This equation can be used to derive q_{st} from the plot of ln $P vs 1/T$, since the slope is equal to—q_{st}/R. Since q_{st} is not entirely independent of T, then a number of accurate experimental isotherms are necessary to evaluate q_{st}. Adsorption isotherms for krypton on a silver iodide sample at temperatures ~78 K are shown in part in Fig. 21.2 and the derived isosteric heat values in Fig. 21.3.

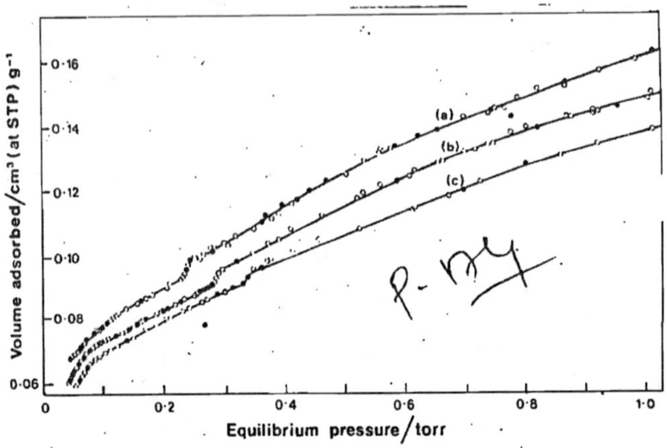

Fig. 21.2: Adsorption isotherms for krypton adsorbed on silver iodide at (a) 77.49, (b) 78.30 and (c) 79.10 K in the pressure range 0.04 -1.00 torr (1 torr = 133.3 Pa).

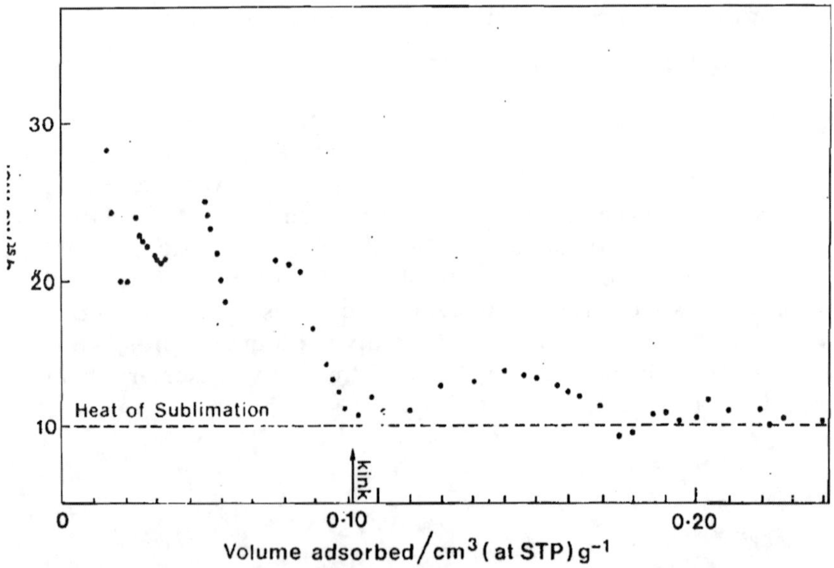

Fig. 21.3: Isosteric heat values as a function of coverage from the data shown in part in Fig. 21.2.

22

Surface Heterogeneity

Introduction

In our treatment of adsorption and of chemical reactions on surfaces we have so far been assuming that all the surface sites are of the same character. The Langmuir adsorption isotherm, for example, is based on this assumption of surface homogeneity. In reality, as H.S. Taylor first pointed out, surfaces are never absolutely smooth, and we have to take account of variations in surface activity.

Various lines of experimental evidence provide evidence for the *inherent heterogeneity* of surfaces. For example, if a metal is heated for a period of time, its capacity for adsorption and catalysis is usually decreased. This is due to sintering of the surface, resulting in a decrease in the number of atoms that constitute the most active centers.

There is also much kinetic evidence for the variability of surfaces. For example, the decomposition of ammonia on molybdenum is retarded by nitrogen, but as the surface becomes saturated by nitrogen, the rate of the decomposition does not fall to zero. This suggests that the reaction can occur on certain surface sites on which the nitrogen cannot be adsorbed.

In view of the fact that surfaces show such variability, it may at first seem surprising that fairly simple kinetic laws often apply to surface-catalysed reactions. The reason is that because of the exponential term in the rate equation a reaction will occur predominantly on the most active surface sites. For example, suppose that a reaction can occur at ordinary temperatures on 10% of the surface sites with an activation energy of 50 kJ mol^{-1} and on the remaining 90% with activation energies ranging from 80 to 150 kJ mol^{-1}. In this situation the amount of reaction occurring on the 90% of less active sites will be negligible compared with that occurring on the 10% of most active sites. The variability of the 90% of the surface will therefore have little effect on the kinetic behaviour.

Aside from an *inherent heterogeneity*, resulting in a true variation in surface sites, there can also be an *induced heterogeneity* resulting from the interactions between adsorbed molecules. Suppose, for example, that there is a significant repulsion between molecules that are adsorbed side by side on a surface. The result will be that as the substance is progressively adsorbed, the first molecules will not be close together, but subsequent molecules will necessarily be close to other molecules and will be adsorbed less strongly. The adsorption behaviour is thus very similar to that occurring when the surface is inherently heterogeneous; the first molecules will be attached more strongly than the later ones.

The heat evolved on chemisorption usually falls as the surface is progressively covered. This can be due to inherent heterogeneity, to induced heterogeneity, or to a combination of the two.

Many surface catalysts are much more efficient if they are *promoted*. A **promoter** is a substance that in itself has little or no catalytic power but which when added to a catalyst greatly improves its performance. A pure iron surface, for example, is not particularly good catalyst for the synthesis of ammonia from nitrogen and hydrogen; however, addition of small amounts of various materials, such as Al_2O_3, brings about a very considerable improvement in the catalytic performance. Promoters bring about their action in a variety of ways. Some of them act mainly by changing the structure of the surface, with an increase in the surface area. Promoters also act by producing surface regions, such as phase boundaries, which have a greater catalytic activity than other parts of the surface.

Most catalysts are easily poisoned by substances that are strongly adsorbed at the active centers on the surface. Since the proportion of active centers may be small, minute amounts of poison can cause a very considerable decrease in the effectiveness of a solid surface.

23

Monolayer Models

Introduction

Before proceeding to the discussion of the relationship between experimental adsorption isotherms and physical models of the adsorption process, it is essential to consider the ways in which adsorption data are presented. Probably because of the way in which early workers in the field, such as Langmuir, represented their data, there has evolved the use of terminology which is proving most difficult to supersede. The simplest way to represent the amount of a particular molecule adsorbed at a particular temperature and equilibrium pressure would be as either the number of moles, n^s, or molecules, N^s, in the adsorbed layer, thus

$$N^s/L = n^s = f(P,T) \tag{1}$$

but it is often expressed instead as the equivalent volume of n^s moles of adsorbate at STP, V^s

$$V^s = 22414\ n^s = f(P,T) \tag{2}$$

where V^s is now in cm^3 at STP. Therefore an adsorption isotherm may be represented as either

$$V^s{}_T = f(P) \tag{3}$$

or

$$n^s{}_T = f(P) \tag{4}$$

It is possible to describe a monolayer in an exactly analogous way to that used for an insoluble monolayer at the air-water interface, and consider what equations of state might be adopted for the adsorbed film. The ideal 2-d gas equation would suggest itself as a starting point,

$$\Pi A = kT \tag{5}$$

and the problem becomes one of relating the parameters which cannot be measured directly, Π and A, to those which can, P and n^s.

We may write the Gibbs adsorption isotherm for the adsorption of a one component gas as

$$-d\gamma = \Gamma \, d\mu \tag{6}$$

which on substitution for Γ yields

$$-d\gamma = \frac{n^s}{A}\, d\mu = \frac{n^s{}_m}{A} \cdot \frac{n^s}{n^s{}_m}\, d\mu \ . \tag{7}$$

However, the quantity $n^s/n^s{}_m$, where $n^s{}_m$ is the amount in a complete monolayer, is the fractional coverage, θ, thus

$$-d\gamma = \frac{n^s{}_m}{A}\, \theta \, d\mu \ . \tag{8}$$

The chemical potentials in the adsorbed and equilibrium gas phase must be equal, therefore

$$\mu = \mu_{gas} = \mu^{\varnothing} + kT \ln P \tag{9}$$

and

$$d\mu = kT \, d(\ln P) \tag{10}$$

Substitution into equation (8) yields

$$-d\gamma = \frac{n^s{}_m}{A}\, kT \, \theta \, d(\ln P) \ . \tag{11}$$

If this equation is integrated with the boundary conditions that $\gamma = \gamma_o$ when P = 0, and the upper limit that the surface tension is γ when the pressure is P, then we may write

$$\int_{\gamma_0}^{\gamma} d\gamma = \frac{n^s_m}{A} kT \int_{o}^{P} \theta \, d(\ln P) \; . \qquad (12)$$

Then by writing $A_m = A/n^S{}_m$, and using the definition of the surface pressure, $\Pi = \gamma_0 - \gamma$, we have

$$\Pi = \frac{kT}{A_m} \int_0^P \theta \, d(\ln P) \; . \qquad (13)$$

The ideal 2-d gas equation (5) may be rewritten

$$\Pi = \frac{kT}{A} = \frac{kT}{A_m} \cdot \frac{A_m}{A} = \frac{kT}{A_m} \quad \theta \qquad (14)$$

Therefore differentiating and using equation (13) we have

$$d\Pi = \frac{kT}{A_m} \, d\theta = \frac{kT}{A_m} \, \theta \, d \, (\ln P) \qquad (15)$$

or

$$d(\ln \theta) = d(\ln P)$$

whence $P = K\theta$ (16)

which is known as Henry's Law, where K is a constant.

The derivation of Henry's Law just given is a particular example of the manipulation of a 2-dimensional equation of state, which may in general be written

$$\Pi = \frac{kT}{A_m} \, f(\theta,T). \qquad (17)$$

This would give

$$\ln P = \int_0^{\theta} f' \, (\theta,T)_T \, d(\ln \theta) + \ln K \qquad (18)$$

where K is an integration constant, and the generalised model isotherm function would be

$$P = K \, g \, (\theta,T) \qquad (19)$$

This has been well presented by Ross and Olivier.

The first improvement upon the ideal equation is to make

allowance for the finite size of the adsorbing species, as in equation (20)

$$\Pi\,(A - A_0) = kT \tag{20}$$

which using the above approach yields the Volmer adsorption isotherm,

$$P = K\frac{\theta}{1-\theta}\exp\left(\frac{\theta}{1-\theta}\right) \tag{21}$$

The further extension of the equation of state to the 2-d analogue of the van der Waals equation

$$\left[\Pi + \frac{\alpha}{A^2}\right](A - \beta) = kT \tag{22}$$

where α and β are contants, gives the adsorption isotherm

$$P = K\frac{\theta}{1-\theta}\exp\left[\frac{\theta}{1-\theta} - \frac{2\alpha\theta}{kT\beta}\right] \tag{23}$$

often known as the Hill-de Boer equation

The constant *K* in equations ((16), (21) and (23) may be related to the adsorption potential, ${}_mU_o$ by

$$K = A^0 \exp\left(-{}_mU_o/RT\right) \tag{24}$$

where

$$\ln A^0 = \frac{\Delta S^a}{R} + \frac{E^{vib} - E_o^{vib}}{RT} + \frac{\Delta E^{tr}}{RT} + \frac{\Delta E^{rot}}{RT} + \frac{\Delta E^{ia}}{RT} - \ln g(\theta,T) + \ln P^{\varnothing} - 1 + \frac{A_m}{R\theta}\left[\frac{\partial\Pi}{\partial T}\right]_\theta \tag{25}$$

In this equation ΔS^a is the integral entropy of adsorption, ΔE^{tr} and ΔE^{rot} are the translational and rotational energy changes on adsorption, ΔE^{ia} is the potential energy of interaction of the adsorbate, and $P^{\varnothing}$ is the standard pressure. Therefore A^o can be evaluated for a particular isotherm function, $g(\theta,T)$.

These three theoretical isotherms are based on essentially the same assumptions as those used in the kinetic energy thoery of 3-d gases, and therefore are descriptions of mobile 2-d adsorbed films. The Henry's sLaw case would only be expected to apply at very low coverages, and as the coverage increases Volmer's equation would apply. However, neither of these two equations predict phase transitions which can be observed in certain cases, as for example the adsorption of krypton on sodium chloride shown in Fig. 23.1. The Hill-de Boer equation, with appropriate values of the constants α and β, does predict such behaviour. The critical point corresponds to $2\alpha/kT\beta = 6.75$, when $\theta = 1/3$ and $P/K = 0.08689$. Above the critical value of $2\alpha/kT\beta$ the euqation has one real and two imaginary roots and therefore predicts a unique value for the amount adsorbed at a particular pressure, but below the critical value of $2\alpha/kT\beta$ there is the possibility of there being 3 real roots, as can be seen in Fig. 23.2. For example the isotherm for $2\alpha/kT\beta = 10$, in fact would follow the solid line, whereas the equation predicts the dashed curve. The phase transition between the point g, where the film is a 2-d gas, and the point 1, where it is a 2-d liquid, coccurs at the pressure where the chemical potentials of the gas and liquid phases are equal.

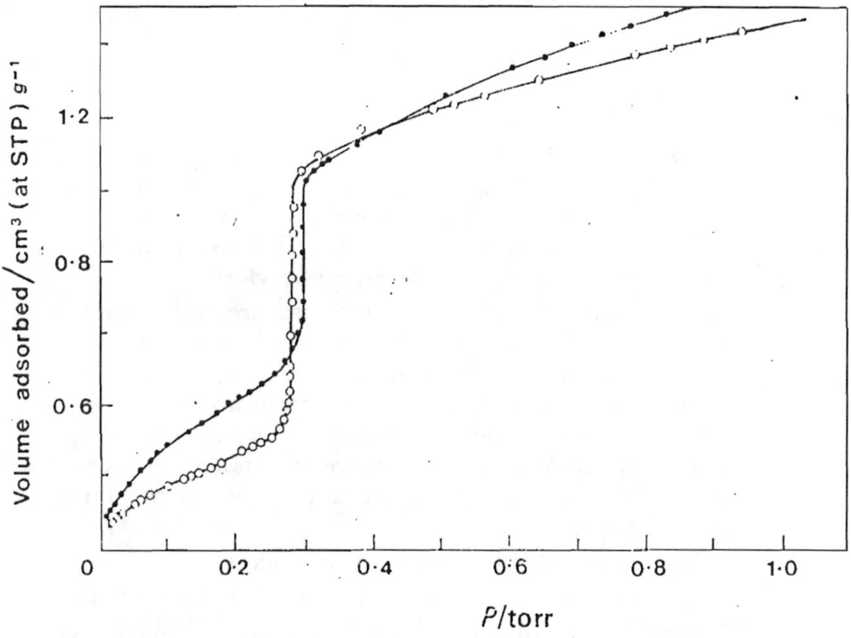

Fig. 23.1: Adsorption isotherms at 76.1 K for krypton on sodium chloride.

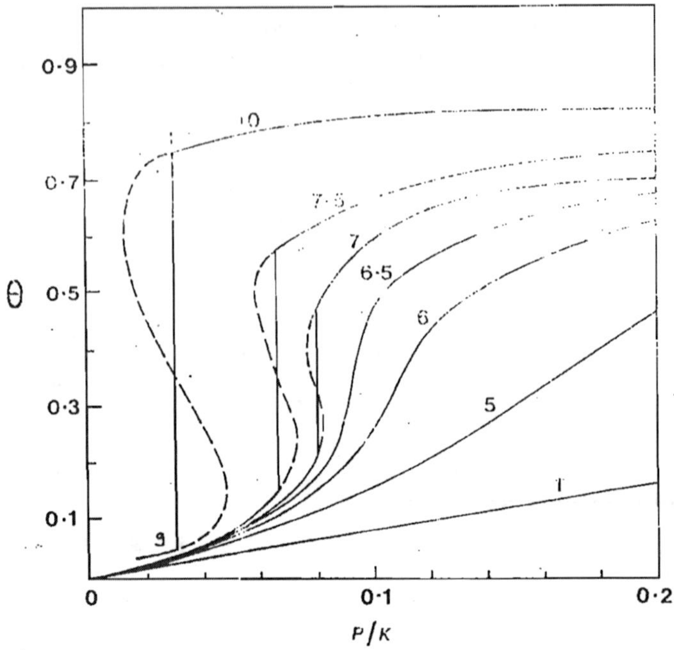

Fig. 23.2: The Hill-de Boer equation (23) plotted for various values of $2\alpha/kT\beta$.

Models for localised monolayers cannot readily be derived from considerations of the corresponding equations of state, although the latter do exist. The first approach was by Langmuir, who used a kinetic derivation. The assumptions upon which it is based have been clearly stated by Fowler.

> "necessary and sufficient assumptions... are that the atoms (or molecules) of the gas are adsorbed as wholes onto definite points of attachment on the surface of the adsorber, that each point of attachment can accommodate one and only one adsorbed atom, and that the energies of the states of any adsorbed atom are independent of the presence or absence of other adsorbed atoms on neighbouring points of attachment".

This defines an ideal localised monolayer.

Langmuir considered the dynamic equilibrium between the adsorbed layer and the bulk gas phase. The kinetic theory of gases gives that the number of molecules striking unit area of surface in unit time as

$$i = P/(2\pi mkT)^{1/2} \tag{26}$$

Adsorption will not occur if a molecule hits a site that is already occupied, and of those molecules hitting a bare site, only a certain fraction, j, will be captured (that is, adsorbed), the remainder being reflected. If the fraction of sites already occupied is θ, then the rate of adsorption will be ij (1-θ). The rate at which molecules return to the gas phase from the adsorbed film can be written as lθ, where l is the rate of evaporation from a fully covered surface. At equilibrium the number of molecules in the adsorbed film must be a constant, therefore

$$ij\,(1-\theta) = l\theta \tag{27}$$

Substituting of i in equation (27) we have

$$\theta = \frac{bP}{1 + bP} \tag{28}$$

or

$$bP = \frac{\theta}{1-\theta} \tag{29}$$

where

$$b = \frac{j}{l(2\pi mkT)^{1/2}} \tag{30}$$

which are representations of the Langmuir adsorption isotherm.

This isotherm has been deduced many times and in many different ways, but the thermodynamic approach leads to an explicit definition of the constant b:

$$b = \frac{h^3}{kT\,(2\pi mkT)^{3/2}} \, \frac{f_a(T)}{f_g(T)} \quad \exp\,(q/kT) \tag{31}$$

where $f_a(T)$ and $f_g(T)$ are the internal partition functions for a molecule in the adsorbed and gaseous state respectively, and q is the energy required to transfer a molecule from the lowest adsorbed state to the lowest gaseous state. It should be noted that it is possible to transpose the Langmuir isotherm into the form of equation (19) where K = 1/b and $g(\theta,T) = \theta/(1-\theta)$.

As was stated earlier in Fowler's description of an ideal localised monolayer, the Langmuir isotherm specifically excludes lateral interaction between adsorbed molecules on neighbouring sites. Obviously this interaction must exist, and a statistical thermodynamic approach leads to

$$P = K \frac{\theta}{1-\theta} \exp -\left[\frac{\varepsilon\theta}{kT}\right] \tag{32}$$

where ε is the lateral energy of interaction. The usual derivation replaces c with $c\omega/2$, where c is an assumed co-ordination number of the underlying lattice and ω is the pair interaction energy for molecules on neighbouring sites. Equation (32) is usually referred to as the Fowler-Guggenheim equation. This form of equation produces isotherms very similar in form to the Hill-de Boer equation, as can be seen in Fig. 23.3. In the Fowler-Guggenheim case the critical condition corresponds to $-\varepsilon/kT = 4, \theta = 0.50$ and $p/k = 0.1353$.

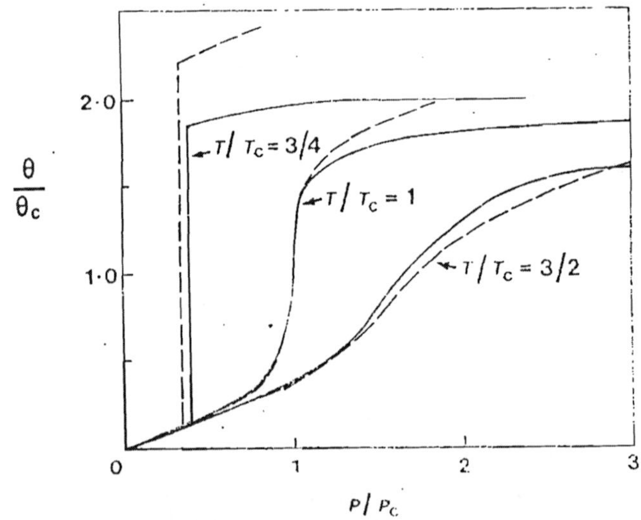

Fig. 23.3: Comparison of theoretical isotherms in reduced units for the Hill-de Boer equation (23) and the Fowler-Guggenheim equation (32). θ_c, T_c, and P_c are the values at the theoretical critical points for 2-d condensation. The values of these parameters are not the same for each model.
- - - - - - Hill-de Boer, ———Fowler-Guggenheim.

The prime limitation of all these models in relation to their comparison with experimentally determined isotherms is that they assume a single value for K, that is the surface is homogeneous or at the very least homotattic. In any real case the surface is heterogeneous, although the overwhelming majority of workers choose to overlook the fact.

24

Multilayer Models

Introduction

From early days of the determination of experimental physical adsorption isotherms, it was realised that adsorption normally did not stop at a monolayer and that multilayers were usually present when the relative pressure, $x = P/P_o$ (P_o is the saturation vapour pressure), is greater than about 0.1 Brunauer noted that there seemed to be five different isotherm types, as illustrated in Fig. 24.1. These shapes are thought to have the following significance:

Type 1 is of the shape predicted by the Langmuir isotherm being characterised by the assymptotic approach to the monolayer volume $V^s{}_m$, where $\theta = V^s/V^s{}_m$;

Type II is the most common type of isotherm corresponding to multilayer formation on a surface of high adsorption potential;

Type III is comparatively uncommon, corresponding to multilayer formation on a solid for which the monolayer adsorption potential is low, being of the same order as that for liquefaction of the adsorptive. Examples have been found for example for bromine on silica gel and nitrogen on ice;

Type IV and V are the analogues of types II and III on porous adsorbents where the adsorption is limited by the voume of mesopores, causing the adsorption to level off at a pressure less than P_0. They reflect condensation phenomena and may show hysteresis effects.

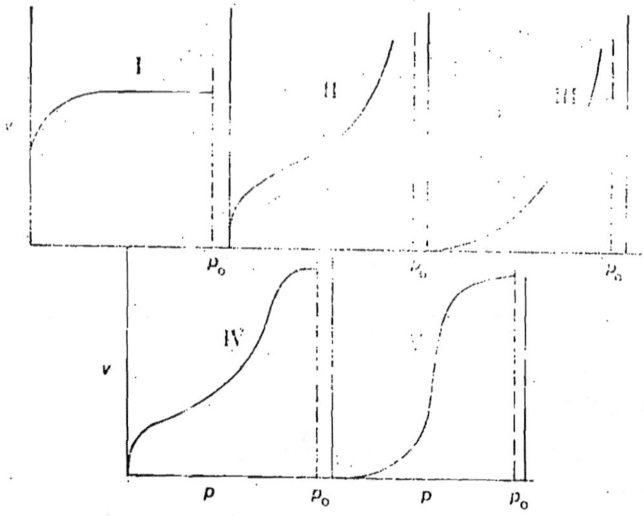

Fig. 24.1. Brunauer's classification of five types of adsorption isotherm.

Although there had been a number of earlier attempts to extend the ideal localised monolayer treatment to include multilayer formation, it was not until 1938 that the Brunauer, Emmett and Teller theory (BET) appeared in the literature. Although subsequently the BET isotherm has also been derived by statistical thermodynamic methods, the original kinetic derivation is still interesting and instructive.

The generalisation of the ideal localised monolayer treatment to include multilayer adsorption is achieved by assuming that each first layer adsorbed molecule serves as a site for the adsorption of a molecule into the second layer, and so on. The concept of localisation remains throughout all the layers and the forces of mutual interaction are neglected, as in the Langmuir isotherm. $S_0, S_1, S_2, S_3 \ldots S_i \ldots$ are defined as the areas covered by, respectively, 0, 1, 2, 3. . . . i . . . molecular layers of adsorbate. At equilibrium the rate of condensation on S_0 is equal to the rate of evaporation from S_1, that is

$$a_1 P S_0 = b_1 S_1 e^{-E_1/RT} \tag{1}$$

where E_1 is that first layer heat of adsorption, assumed constant, and where a_1 and b_1 are constants. Equation (1) is formally identical with the simple Langmuir equation. Similarly the equilibrium between the first and second molecules is expressed by

$$a_2PS_1 = b_2S_2e^{-E_2/RT} \tag{2}$$

and so on, the general statement of the equilibrium between the $(i-1)$th and ith layers being

$$a_1PS_{i-1} = b_iS_ie^{-E_i/RT} \tag{3}$$

The total surface area of the solid is given by

$$A = \sum_{i=o}^{i=\infty} S_i \tag{4}$$

and the total volume of adsorbed gas by

$$V^s = V^s{}_o \sum_{i=o}^{i=\infty} iS_i \tag{5}$$

V_o^s being the volume adsorbed per unit surface when the monolayer is completely filled. Dividing equation (5) by equation (4) gives

$$\frac{V^s}{AV_o^s} = \frac{V^s}{V_m^s} = \sum_{i=o}^{i=\infty} iS_i \Big/ \sum_{i=o}^{i=\infty} S_i \tag{6}$$

where $V^s{}_m$ is the adsorbed volume in a complete ideal localised monolayer. Baly had earlier arrived at what is essentially equation (6) but was able to proceed further only by empirical means. The key to the BET solution lies in the simplifying assumptions that

$$E_2 = E_3 = \ldots . E_i = E_L \tag{7}$$

where E_L is the heat of liquefaction of the bulk liquid, and that

$$\frac{b_2}{a_2} = \frac{b_3}{a_3} \ldots \frac{b_i}{a_i} \tag{8}$$

In other words, the evaporation-condensation properties of the second and higher layers are assumed to be the same as those of the surface of the bulk liquid.

Equation (1) may be written

$$S_1 = yS_o$$

where

$$y = (a_1/b_1)Pe^{E_1/RT} \tag{9}$$

Similarly, using equations (7) and (8) equation (2) may be written

$$S_2 = xS_1$$

where

$$x = (a_2/b_2)Pe^{E_L/RT} \tag{10}$$

and in the general case

$$S_i = xS_{i-1} = x^{i-1} S_1 = yx^{i-1} S_o = Cx^i S_o \tag{11}$$

where

$$C = \frac{y}{x} = \frac{a_1 b_2}{a_2 b_1} e^{(E_1 - E_L)/RT} \tag{12}$$

Substitution of equation (2) in (6) yields

$$\frac{V_s}{V^s_m} = \frac{CS_0 \sum_{i=1}^{i=\infty} ix^i}{S_0 \{1 + C \sum_{i=1}^{i=\infty} x^i\}} \tag{13}$$

The summation in the denominator is the sum of an infinite geometric progression given by

$$\sum_{i=1}^{i=\infty} x^i = \frac{x}{1 = x} \tag{14}$$

while that in the numerator may be transformed thus:

$$\sum_{i=1}^{i=\infty} ix^i = x \frac{d}{dx} \sum_{i=1}^{i=\infty} x^i = \frac{x}{(1-x)^2} \tag{15}$$

Substituting these results in equation (13) gives

$$\frac{V^s}{V^s_m} = \frac{Cx}{(1-x) \quad (1 - x + Cx)} \tag{16}$$

On a free surface the amount adsorbed at saturation is infinite. Consequently at $P = P_o$, in order to make $V^s = \infty$ x, in equation (16) must equal unity. From equation (10) therefore

$$(a_2/b_2)P_o e^{E_L/RT} = 1 \quad (17)$$

when

$$x = P/P_o \quad (18)$$

Substitution in euqation (16) yields

$$V^s = \frac{V^s_m CP}{(P_o - P)\{1 + (C - 1) P/P_o\}} \quad (19)$$

which is known as the 'simple' or '∞-form' BET equation. This equation may be transposed to the linear form

$$\frac{x}{V^s(1-x)} = \frac{1}{V^s_m C} + \frac{C - 1}{V^s_m} x \quad (20)$$

Thus a plot of $x/V^s(1-x)$ against x should give a straight line having slope $(C - 1) V^s_m C)$ and intercept $(1/V^s_m C)$. If the additional assumption is made that

$$\frac{a_1 b_2}{a_2 b_1} \approx 1 \quad (21)$$

the value of C is given by

$$C = e^{(E_1 - E_L)/RT} \quad (22)$$

The term $E_1 - E_L$, the difference between the heat of adsorption in the first layer and the heat of liquefaction, is known as the "net' heat of adsorption. Thus the simple BET equation provides in terms of the model an estimate of both the heat of adsorption and the surface area of the solid.

If, owing to lack of space, adsorption at saturation is restricted to n layers, the BET methods lead to the isotherm

$$V^s = \frac{V^s_m Cx}{(1-x)} \frac{(1 - (n + 1)x^n + nx^{n+1})}{(1 + (C - 1)x - x^{n+1})} \quad (23)$$

where the quantities x, C and V^s_m have the same meanings as previously. Equation (23), which describes adsorption in a limited space such as a capillary, is commonly known as the 'n-layers' BET equation. This equation is of interest for two main reasons. Firstly it may be regarded as an empirical three constant isotherm equation, which has in the past proved useful in the evaluation of surface areas, and secondly, when n =1 it reduces at all values of C (unlike the simple BET equation (37) which only reduces to this form when $x \ll 1$ and $C >> 1$) to the Langmuir equation

$$V^s = \frac{V^s_m \, CP/P_o}{1 + CP/P_o} \tag{24}$$

The simple BET equation (19) is capable of describing more or less adequately most Type II and Type III isotherms, the type being a function of the value of the constant C, but in general only type II isotherms corresponding to high C values yield reliable values of V^s_m.

The mathematical character of the BET equation can perhaps be more readily seen if it is rearranged in the form

$$\frac{V^s}{V^s_m} = \frac{1}{1 - x} - \frac{1}{1 + (C - 1)x} \tag{25}$$

where $x = P/P_0$. Thus the adsorption isotherm of V^s/V^s_m against x consists of the difference between the upper branches of two rectangular hyperbolae representing the first and second terms on the right-hand side of equation (25). The first has asymptotes at $V^s/V^s_m = 0$ and $x = 1$, while the second has asymptotes at $V^s/V^s_m = 0$ and $x = -1/(C-1)$, and also cuts the V^s/V^s_m axis at $V_s/V^s_m = 1$. The difference between these two curves gives the BET isotherm, the shape of which is strongly dependent on C. For C values greater than 2 an isotherm of Type II results, whereas if C is less than 2 the resulting isotherm is of Type III. The point of inflection on the Type II curve is not necessarily at a value of $V^s/V^s_m = 1$ as has been often stated. It can be shown that this is the case when $C = 0$ and when $C \rightarrow \infty$. However, the discrepancy is only marked when $C < 9$. The effect of varying the constant C on the shape of the isotherm can be seen in Fig. 24.2.

There are certain problems associated with the application of the BET isotherm to experimental situations. Firstly it is usually found only to agree with experimental isotherms for non-porous heterogeneous surfaces in the pressure range $0.05 < x < 0.35$, whereas the model assumes a homotattic or homogeneous surface. The multilayer isotherm on such a uniform surface would be expected to be a series of steps, corresponding to the adsorption of successive layers, and these have been observed in practice; for example the adsorption of argon and krypton on graphitised P-33 carbon black at 77 K. They are also partially visible for the adsorption of krypton on silver iodide referred to earlier, which can be thought of as a heterogeneous sample with some homotattic patches.

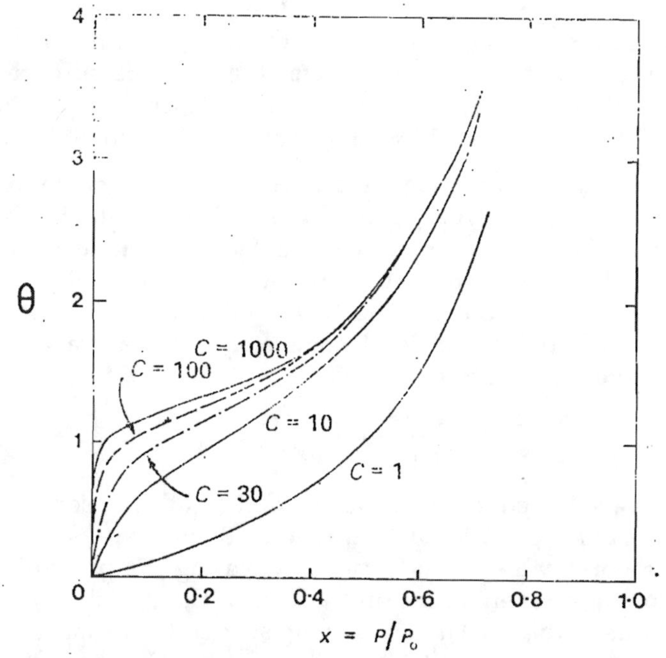

Fig. 24.2: Theoretical BET isotherms (equation (19)) for various values of the constant C.

A number of attempts have been made to improve upon the BET 'simple' equation. Many of these are not well founded, but some still have importance. Anderson assumed that the heat of adsorption in "the second to about the ninth layer" differs from E_L by a constant amount. This gives

$$\theta = \frac{C\,kx}{(1 - kx)\,(1 + (C - 1)\,kx} \qquad (26)$$

which is of the same form as (19) but with x multiplied by a constant, k. A second modification allowed for the situation, likely to be found with porous solids, where the area available for adsorption decreases with successive layers. If the available area in the nth layer is j times that in the (n-1)th layer, the equation is

$$\theta = \frac{Cx}{(1-jx)(1+(c-j)x)} \tag{27}$$

As would be expected, increasing the number of parameters in the equations increases the range of relative pressure where the equations can be made to concur with experiment, but they give $V^s{}_m$ values about 10% higher than the BET equation(19).

Again using the ideal localised monolayer model on a homotattic surface, Huttig deduced a multilayer isotherm which assumes that the evaporation of a molecule from the ith layer is entirely unaffected by the presence of a molecule in the (i + 1)th layer, whereas the BET model assumes that the presence of an (i + 1)th molecule completely prevents evaporation from layers beneath it. This yields

$$\frac{x(1+x)}{V^s} = \frac{1}{CV^s{}_m} + \frac{1}{V^s{}_m}x \tag{28}$$

Although this equation often fits well (ethane adsorbed on sodium chloride at 90 K fits equation (28) where $0.1 < x < 0.7$) it is theoretically badly founded, violating the principle of microscopic reversibility, and allowing the possibility of a molecule in the (i + 1)th layer being suspended in space with no molecule beneath it in the ith layer. The values of $V^s{}_m$ obtained by the Huttig equation are usually larger than those obtained by the BET method by an amount dependent on the C constant, but usually less than 20% greater.

An alternative approach to multimolecular adsorption was adopted by Harkins and Jura, in which they proposed the adoption of the euqation of state for the adsorbed film

$$A = b - a\Pi \tag{29}$$

where a and b are contants, which is often an adequate representation of the liquid-condensed state of insoluble uncharged monolayers at the air-water interface. It would seem reasonable that a multilayer film of adsorbed gas could have similar properties. Applying the Gibbs transformation gives

$$\ln x = B - (A / V^s)^2) \tag{30}$$

The specific surface area, s, may be related to the constant A by

$$s = k_{HJ} A^{1/2} \tag{31}$$

where k_{HJ} is a constant which has to be determined by the independent determination of s.

What is in many ways a very attractive method of visualising multilayer adsorption is to examine the parallel between the distribution of the gas molecules of the earth's atmosphere in the earth's gravitational field, and adsorbed gas molecules at the surface of a solid in its adsorption potential field Fig. 24.3. These ideas were first set down by Polayni who envisaged equipotential surfaces, the distance between which correspond to definite volumes, and that there will be a relationship between the potential ε and the volume ϕ. Thus the adsorption process may be represented by the function $\varepsilon = f(\phi)$, which is a distribution function. Since the potential is independent of temperature, then $\varepsilon = f(\phi)$ represents a 'characteristic curve' for adsorption at all temperatures.

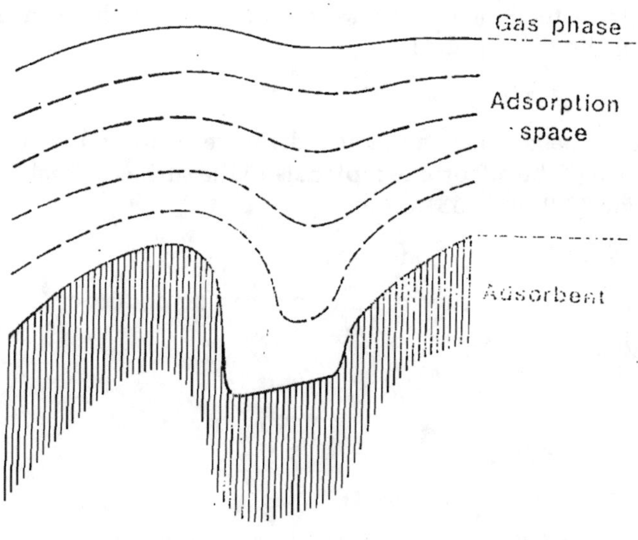

Fig. 24.3: Cross-section of the adsorbed layer showing equipotential contours according to potential theory.

Polanyi postulated three separate cases describing the state of the adsorbed film:

Case 1 At temperatures well below the critical temperature of the adsorbate, the adsorbed film will be a liquid;

Case II At temperatures only just below the critical temperature the adsorbed layer will be a mixture of liquid and compressed gas;

Case III At temperatures above the critical tempeature the adsorbed phase is a compressed gas.

Berenyi elaborated the treatment of Case I by assuming that the gas phase is ideal, the liquid adsorbed film is incompressible, and that a negligible amount of work is involved in forming the liquid surface. He showed that the potential at a distance x from the surface is given by

$$\varepsilon_x = RT \ln(P_o/P) \tag{32}$$

where P is the equilibrium pressure, as previously.

Considering multilayer adsorption only, if we make the assumption that only dispersion forces are responsible for adsorption, then the potential should decrease with the inverse cube of the distance. Then

$$\varepsilon_x = \varepsilon_o/(a + x)^3 \tag{33}$$

where ε_o is a constant and a is a distance defining the closest approach of the adsorbate molecule to the surface. Combining equation (32) and (33) gives

$$RT \ln (P_o/P) = \varepsilon_o/(a + x)^3 \tag{34}$$

from which the isotherm equation may be deduced

$$V^s = -\alpha + \beta\omega^{-1/3}, \tag{35}$$

where

$$\alpha = a\, v_o\, S/v_1$$

$$\beta = (\alpha/a)(\varepsilon_o/RT)^{1/3} \text{ and}$$

$$\omega = \ln(P_o/P)\,,$$

in which v_0 is the molar volume at STP and v_l is the molar volume of the liquid.The chief limitations of equation (35) are likely to be that at low values of P/P_0 wher the adsorbed film is thin, then the assumption that the solid is continuous, which is the basis of equation (33), is likely to be a poor one. Furthermore if on ionic solids the electric field contribution to the adsorption potential is appreciable then equation (33) will once again be inappropriate. This approach which is due to Frenkel, Halsey and Hill, is more commonly quoted in the form where a = 0, which gives

$$\theta^n = A_{FHH}/\ln (P_0/P) \qquad (36)$$

where

$$A_{FHH} = \varepsilon_o (x_m)^n RT$$

and x_m is the adsorbed film thickness at the monolayer point. This equaiton is frequently called the Frenkel-Halsey-Hill equation, and their theory referred to as Slab Theory. Experimental values of n, which should be 3 if van der Waals dispersion forces alone are responsible for adsorption, are frequently less than 3, which is interesting since a value of n = 2 gives a form of equation equivalent to the Harkins-Jura equation (30)

An alternative form for ε_x was proposed by de Boer and Zwikker based on the assumption that a polar adsorbent surface could induce dipoles in non-polar adsorbate molecules in the first layer, and that these in turn could induce dipoles in the second layer, and so on. This would predict the form of equation

$$\varepsilon_x = \varepsilon_o \exp (- \chi^x) \qquad (37)$$

where

$$\chi = (1/a_o) \ln (\alpha/\acute{\alpha}^3)^2,$$

in which a is the polarisability, a_o is the atomic diameter, and a the 'effective' value of distance from the surface, which is smaller than a_o. Combining this with eqution (32) and solving as before gives the equation

$$\ln \ln (P_0/P) = \ln (\varepsilon_0/RT) - (X^{v/v_0}S)V^s \qquad (38)$$

which is a suitable linear form for the isotherm.

It is important to realise that all the above multilayer models assume a uniform surface, whereas all surfaces are to a degree heterogeneous, and most are very heterogeneous. However, as we shall see in the next section, this has not prevented the application of these models to experimental isotherms on heterogeneous surfaces. The only significant attmept to tackle this problem is that due to Halsey, who extended the application of equation (35) to heterogeneous surfaces, with $n = 3$, using a 'patchwise' model. The method is based on distribution function $N_1(V_o)$, giving for the coverage in the mth layer

$$\theta_m = \frac{1}{m^3} \int_{-m^3 RT \ln (P/P_o)}^{\infty} N_1 (V_o) \, d U_o \qquad (39)$$

which leads to the isotherm equation

$$\theta = \exp (U_o/\bar{U}_o) \sum_{m=1}^{\infty} z^{m^3} \qquad (40)$$

where $z = (P/P_o)^{RT/U_o}$, and $\bar{U}_o$ is the modulus of the distribution. Although this equation is difficult to apply to experimental isotherms it does predict a stepped isotherm for a uniform surface, with the gradual disappearance of the steps as the degree of heterogeneity increases.

25

Surface Area Determination

Introduction

In the earlier chapter, we discussed the various models for the multilayer adsorption processes. Firstly therefore we should examine the applicability of these to experimental isotherms, and see what sort of 'goodness of fit' we might expect. If we use readily accessible data, then the reader can extend the process further if he wishes. Data which has been previously used for this purpose is that of Keenan and Holmes who examined the adsorption of nitrogen, argon and oxygen at several temperatures on potassium chloride. We will use two of the most extensive isotherms, namely nitrogen at 78K and argon at 83K.

These two isotherms are both BET type II, which as we have said before is the most common multilayer form, and are shown in Fig. 25.1. These two curves have the appearance of one with a fairly high C constant in the case of N_2, and one with a low C constant for the Ar isotherm. On the latter isotherm the linear portion at low pressure which does not extrapolate back to the origin looks slightly abnormal; a linear portion below the pressure range of these data would correspond to Henry's Law. Argon and nitrogen are the two most frequently recommended gases for surface area determination.

Let us examine the application of the BET 'simple' model to the data. The conventional BET plot is shown in Fig. 25.2. The N_2 data gives a linear plot over the approximate range 0.03

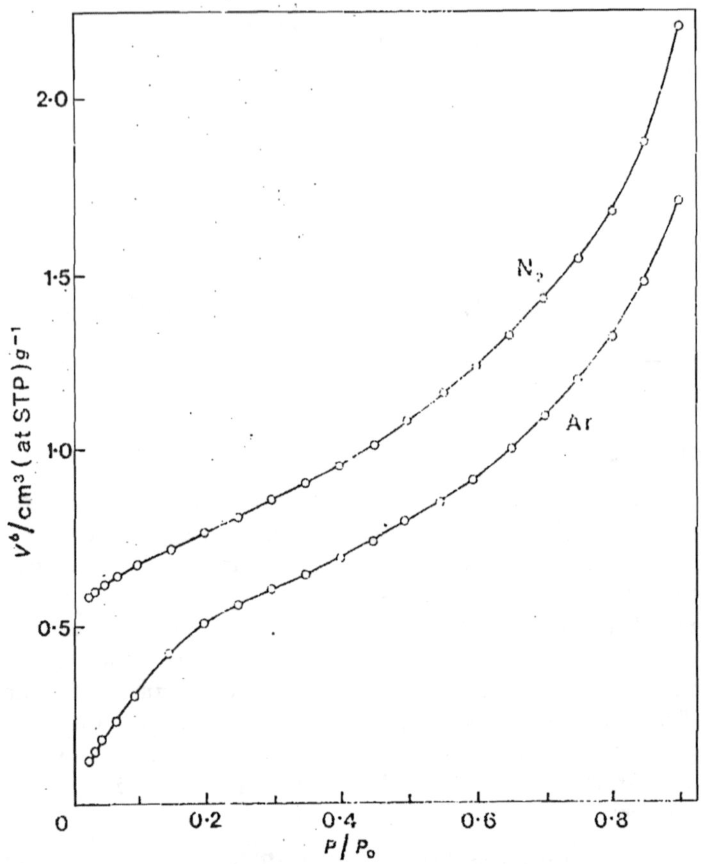

Fig. 25.1: Adsorption isotherms of nitrogen at 78 K and argon at 83 K on potassium chloride. Data of Keenan and Holmes.

$\leq x \leq 0.25$, whereas the linear region of the Ar data is difficult to see, but the range $0.07 \leq x \leq 0.15$ does seem to be linear. Since both isotherms were determined on the same sample then one would like to find that the surface area calculated from both plots is the same.

From the linear region a value of $V^s{}_m$ can be obtained from

$$V^s{}_m = \frac{1}{\text{slope} + \text{intercept}} = \frac{1}{\dfrac{C-1}{CV^s{}_m} + \dfrac{1}{CV^s{}_m}} \qquad (1)$$

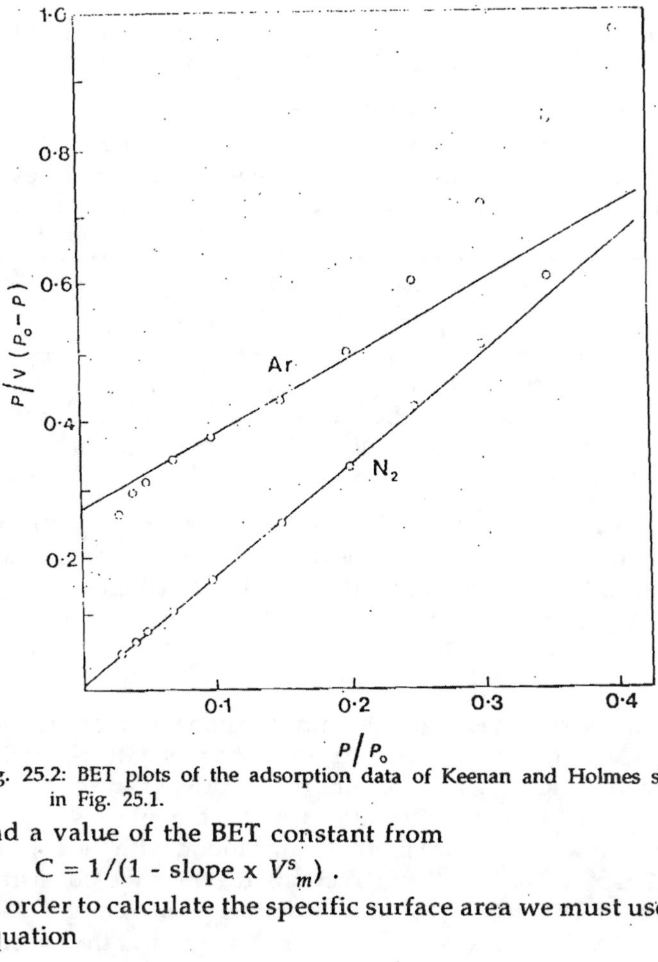

Fig. 25.2: BET plots of the adsorption data of Keenan and Holmes shown in Fig. 25.1.

and a value of the BET constant from

$$C = 1/(1 - \text{slope} \times V^s{}_m) \, . \tag{2}$$

In order to calculate the specific surface area we must use the equation

$$s = \frac{V^s{}_m L \, \sigma_m}{v_o} \quad m^2 g^{-1}$$

$$= 26.87 \, \sigma_m \, V^s{}_m \;\; m^2 \, g^{-1} \tag{3}$$

if $V^s{}_m$ is in cm^2 at STP g^{-1} and σ_m in nm^2 $molec^{-1}$, and v_o is the molar volume. The quantity σ_m is known as the 'molecular cross-sectional area of the adsorbate' or better as the 'adsorption cross-section', and is the area occupied by an adsorbate molecule in a complete monolayer. It is usually calculated from the liquid density of the adsorbate at the adsorption temperature and may be taken as 0.162 nm^2 $molec^{-1}$ and 0.138 nm^2

molec^{-1} for N_2 and Ar respectively. The results of the BET analyses are shown in Table 25.1.

TABLE 25.1

BET analysis of the plots in Fig. 25.2 of the isotherm data shown in Fig. 25.1 of Keenan and Holmes

	Slope	*Intercept*	V^s_m/*cm*^{3} *at STP g*$^{-1}$	C	σ_m/*nm*^{2} *molec*$^{-1}$	*S/m*^{2}*g*$^{-1}$
N_2 at 78K	1.620	0.006	0.615	270	0.162	2.68
Ar at 83K	1.108	0.262	0.730	5.2	0.138	2.71

The values shown in this table indicate close agreement in the value of the specific surface area, and the expected diversity of C constant values. Thus despite the problems, consistent results are obtained.

The modifications to BET theory present various problems. Firstly the Anderson equations do not have readily plottable forms, unless it is assumed that C is high, when

$$x = \frac{1}{K}\left[1 - \frac{V^s{}_m}{V^s}\right] \tag{4}$$

It would be expected that this simplification might be justified for the N_2 data but not for the Ar data. The plot of this equation is shown in Fig. 25.3, and contains the surprise of an apparently quite good fit for the Ar data over the range $0.3 \leq x \leq 0.9$ compared with the more limited range found with the N_2 data of $0.2 \leq x \leq 0.6$. The linear sections do not give consistent surface areas since $^sN_2 = 3.05\ m^2g^{-1}$ and $s_{Ar} = 2.54\ m^2g^{-1}$. Thus the least valid case gives the longer fit range and the specific surface area nearer the BET answer, which offers another example of the fact that an apprent 'fit' of the data does not prove the validity of the model.

The Huttig equation is another example of reasonable 'fits' (Fig. 25.4) giving inconsistent results, and since the model upon which it is based is erroneous, nothing much can be gained from the answers that $^s{}_{N_2} = 2.85\ m^2g^{-1}$ and $s_{Ar} = 2.18\ m^2\,g^{-1}$.

Plotting the data using the axes of the linear form of the Harkins-Jura equation gives the graphs shown in Fig. 25.5.

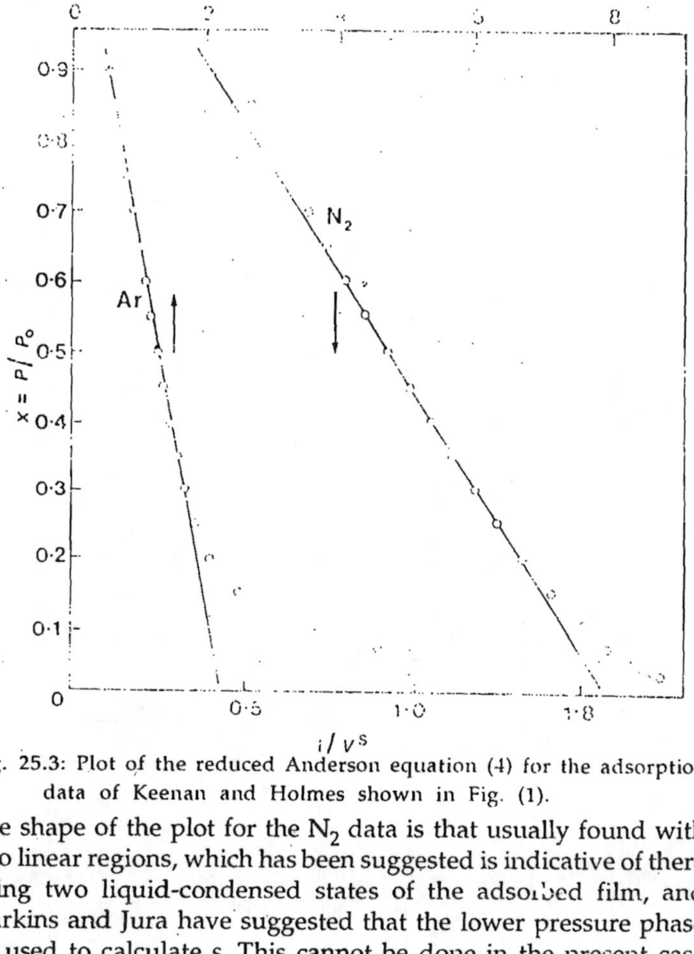

Fig. 25.3: Plot of the reduced Anderson equation (4) for the adsorption data of Keenan and Holmes shown in Fig. (1).

The shape of the plot for the N_2 data is that usually found with two linear regions, which has been suggested is indicative of there being two liquid-condensed states of the adsorbed film, and Harkins and Jura have suggested that the lower pressure phase be used to calculate s. This cannot be done in the present case because of the absence of reference data on a standard potassium chloride sample which would be needed to evaluate the constants. The situation can be even more complicated since it has been suggested that there may be four different liquid phases in the case of N_2 on carbon black.

The argon data in Fig. 25.5 show only one linear region, and would tend to suggest that a liquid condensed state model in this case is only applicable when coverages are greater than about a monolayer. All the N_2 data presented would be in this region. The low pressure N_2 data given by Harkins and Jura also tend to curve in a manner similar to the present Ar data.

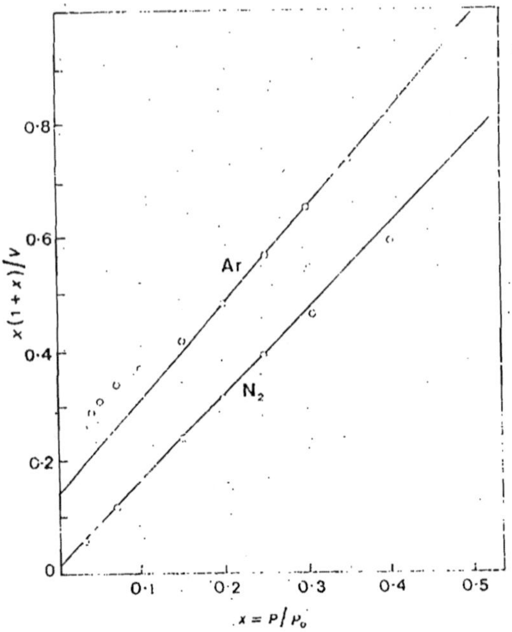

Fig. 25.4: Huttig plots equation of the adsorption data of Keenan and Holmes shown in Fig. 25.1.

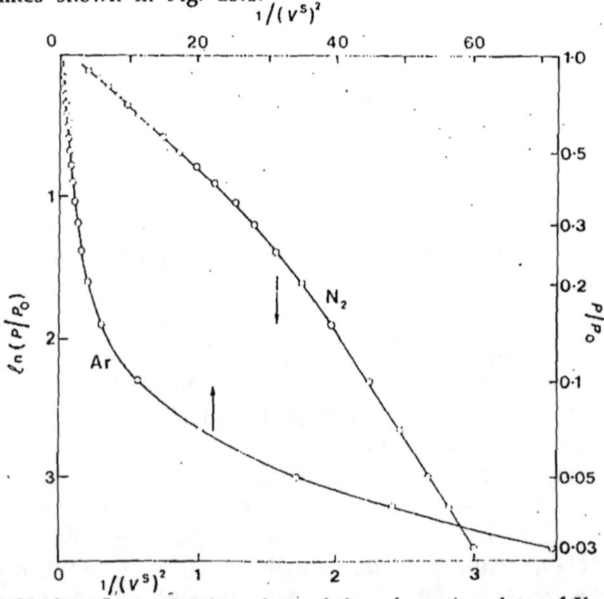

Fig. 25.5: Harkins-Jura equation plots of the adsorption data of Keenan and Holmes shown in Fig. 25.1.

The remaining two models worth discussing are potential theory using equation and plotted in Fig. 25.6 and polarisation theory, and plotted in Fig. 25.7. Both models represent the data reasonably well, but potential theory is the more successful in this respect. However, there are serious difficulties in evaluating the constants to permit their use in the estimation of surface area although the models could be used in a relative fashion, as was the case with the Harkins-Jura equation.

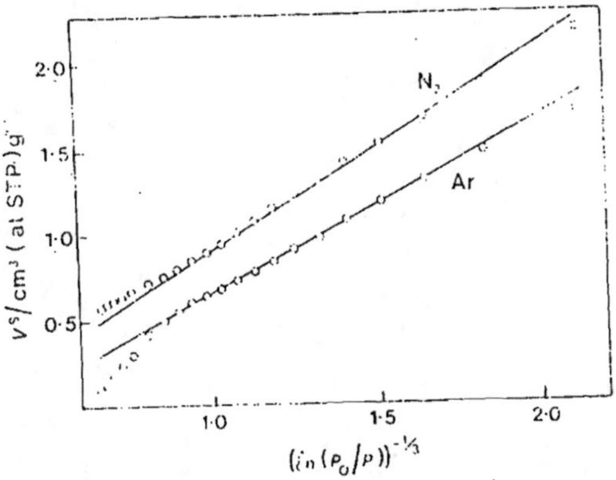

Fig. 25.6: Potential theory plots using of the adsorption data of Keenan and Holmes shown in Fig. 25.1.

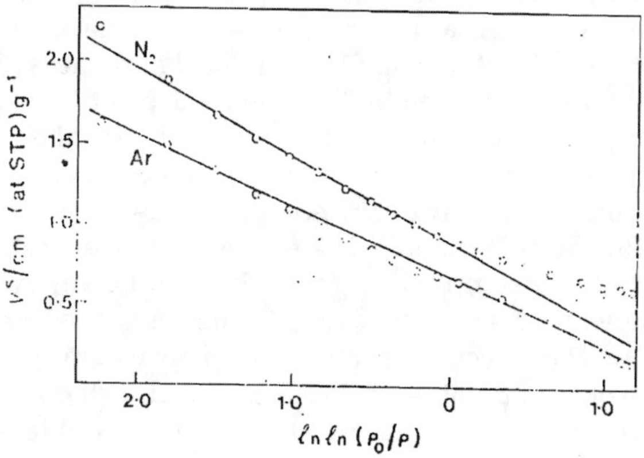

Fig. 25.7: Polarization theory plots using equation of the adsorption data of Keenan and Holmes shown in Fig. 25.1.

The above example highlights the fact that the BET 'simple' equation offers the most practically useful method of evaluating surface areas. The remainder of this section will therefore consider the three gases most frequently used in this connection namely N_2, Ar, and for small specific surface areas Kr.

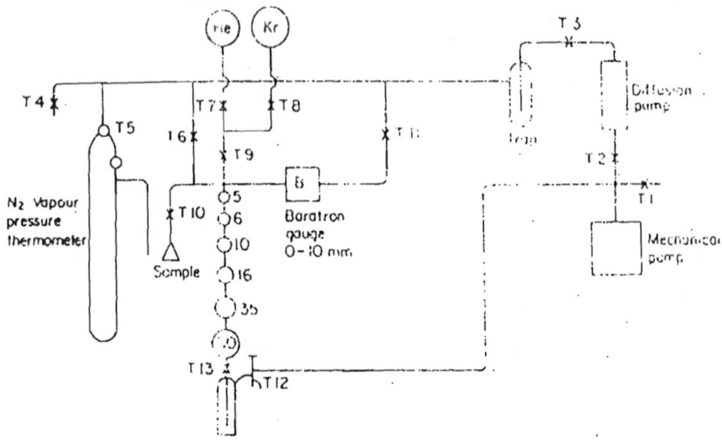

Fig. 25.8: Schematic diagram of a design for a krypton BET volumetric adsorption apparatus.

The techniques of measuring the amount of gas adsorbed at a particular equilibrium pressure are usually volumetric and have been reviewed quite frequently, and the schematic diagram of an apparatus for krypton BET measurements is shown in Fig. 25.8. Although the standard apparatus has been proposed for the nitrogen BET method, the problem is that the actual design is in some measure a matter of personal preference, and also frequently a function of the amount of money available as well as the availability of technological innovations. However, it is useful for anyone entering the field for the first time to be able to see a sample design, which can be varied to suit circumstances, provided that simple calcultions are performed to check on the effects of such changes on accuracy. Because the design of an apparatus for the krypton BET method illustrates most of the problems that are likely to be encountered with any gas, this is the type illustrated in Fig. 25.8, and which will now be discussed in more detail.

Fig. 25.8 shows a schematic diagram of a design based on the following assumptions: *(i)* that the sample has an area 0.1 m^2, *(ii)* that $C \sim 35$, and *(iii)* that the minimum volume of the system is not greater than 15 cm^3. This volume is the sum of the dead space, the tube volume between T_{10}, T_9 and the Baratron, and the Baratron gauge itself. These assumptions mean that only when they are complied with will the apparatus give 6 points, spread throughout the BET range, from one dose of krypton. Deviations from these assumptions may result in a reduction of the number of points with in the range, or a less even spread, but for many samples they will prove satisfactory. Clearly the design represents a compromise which the authors believe to be satisfactory.

There are a number of specific experimental problems which should be understood and which will be dealt with individually.

(a) Pressure-measuring devices. The design shows a 0-10 torr Baratron diaphragm pressure gauge which is quick to use, and is sufficiently precise to make pressure measurement an insignificant source of error. Simpler devices such as McLeod gauges or thermistor gauges can be used, but they result in a lose of accuracy in pressure measurement. The use of a McLeod gauge with a system such as that in Fig. 25.8 would necessitate changes in the gas burette bulb sizes, since the minimum system volume is likely to be ~ 150 cm^3.

(b) Dead space calibration. This is normally done by calibration with helium and results in an equivalent volume, V, at room temperature, T_R. This volume is dependent on T_R.

Let us suppose that V arises from two components, a volume, V_R, above the liquid nitrogen level at T_R, and an actual physical volume, V_N, below liquid nitrogen level at liquid nitrogen temperature, T_N. The total number of moles in the system may then be written

$$n = \frac{PV}{RT_R} = \frac{PV_R}{RT_R} + \frac{PV_N}{RT_N} \qquad (5)$$

Therefore

$$V = V_R + V_N \frac{T_R}{T_N} \tag{6}$$

It is essential, therefore, either to know the values of V_R and V_N or to only use a value of V appropriate to the temperature at the time adsorption measurements are made. Normally errors of this type are small since T_R does not usually change greatly, but, for example, intermittent sunshine can cause problems. Errors due to too large a dead space have been discussed by Ross and Olivier. In general it is undesirable that the dead space should be larger than the burette volume.

(c) *Effect of light*. It has been noted by Rosenberg and Martel that small amounts of light falling on the sample can result in a significant rise of the sample temperature. The sample should therefore always be protected from light and other radiation. Cover glass sample tubes with copper foil, that has been blackened with yellow ammounium sulphide, up to the liquid nitrogen level, and cover the tube with black paint from this junction to beyond the tap isolating the sample from the rest of the system. The copper foil should not extend above the liquid nitrogen level since it would spoil the sharpness of the thermal junction.

(d) Thermal transpiration. This effect causes a reduction of the actual pressure compared to that measured on the gauge, at the lower end of the pressure range. An idea of the errors involved can be seen in the work of Ross and Olivier. Using a 6 mm internal diameter sample tube, they suggest the error would be ~ 2% at P = 0.12 torr or P/P_0 = 0.05, and that it decreases as the sample pressure is increased. The effect would be greater if the sample tube diameter at the junction were less than 6 mm. A 2mm diameter tube would result in an increase in the error to about 10% at the same pressure. Calculations should normally be performed to see whether the error is significant, and the most suitable recent data for this purpose are those of House and Jaycock.

If corrections are to be made then the following points need to be considered. Firstly the equilibrium pressure will be the corrected one, and secondly the correct volumetric estimation of the amount adsorbed involves recognising that there are two pressures in the dead space volume, and that the number of moles of gas present will be given by

$$n = \frac{P_R V_R}{RT_R} + \frac{P_N V_N}{RT_N} \qquad (7)$$

where P_R and P_N are the pressures in the parts of the system at T_R and T_N respectively, and V_R and V_N need to be known. These two volumes may be calculated from measurements of the 'dead space' first with liquid nitrogen present, and then with the whole system at room temperature.

(e) Measurement and control of sample temperature. Sample temperature can best be measured by means of a nitrogen saturation vapour pressure thermometer using standard vapour pressure data. This permits the bath temperature to be measured to at least 0.01 K. The temperature can be controlled by an expensive cryostat, but can be simply done using mixutures of liquid nitrogen and liquid air. If a pressure of about 770 torr on the vapour pressure thermometer is chosen (~77.4 K) then the bath temperature canbe fixed at this by adding small amounts of liquid nitrogen or air to allow for atmospheric pressure variation. The system should be stirred using a slow-speed electric motor to ensure uniformity of temperature, and the liquid nigrogen/air level on the sample tube should be kept constant.

There is an alternative method for observing the temperature, although it does not permit continuous control, and only offers a single measurement at the end of the adsorption mesurements. After the adsorption measurements are complete a large dose of krypton can be admitted to the system and the saturation vapour pressure of solid krypton measured. This may be used to calculate the temperature using the vapour pressure data of Keesom *et al.* It is preferable to use a nitrogen vapour pressure thermometer.

(f) The non-ideality of krypton. Calculations show that with pressures less than 2 torr no correction is necessary for non-ideality, the correction terms being (1-0.00229 P) at room temperature

and (1-0.176 P) at liquid nitrogen temperature, where P is in atmospheres. However, corrections may be significant if a McLeod gauge is used to measure the pressure, but in the system in equilibrium with the solid the error in neglecting non-ideality correction would be a maximum of 0.04% at 77.5 K and 0.0005% at room temperature.

The above considerations are all relevant to N_2 and Ar measurements with the exception of thermal transpiration corrections, which are not necessary in conventional appratus in the pressure ranges where x > 0.03.

If Kr is used there also remains the problem of the approapriate value of Po. The adsorbed film in fact appears to behave more like a liquid than a solid and therefore a vapour pressure equation for the liquid state needs to be extrapolated, since Kr is a solid at liquid N_2 temperatures.

Inorder to convert $V^s{}_m$ values into specific surface areas, then a value for σ_m has to be known. The values usually quoted are calculated from liquid densities, and therefore assume that the adsorbed molecules pack in a similar way on the surface. The fact that the adsorbed film is a liquid is not surprising from the point of view of potential theory, but the hindered translation model calculations discussed earlier would suggest that some degree of localisation will be met with quite frequently, and that the effective value of σ_m for the first layer will be larger than that calculated from the liquid density in this case. This point may be clearly seen in relation to the values σ_m frequently employed in the calculation of surface areas by the krypton BET method, and some are listed in Table 25.2.

This table clearly shows that the necessary experimental values of σ_m are significantly larger than those predicted from either liquid or solid densities, which is consistent with some degree of localisation of the adsorbed film.

The BET equation is normally applicable in the relative pressure range between 0.05 and 0.35, or some slightly more restricted range within those limits. It is often stated that this corresponds to the pressure range where the second layer of adsorbed molecules is being built up on the surface. In many cases krypton BET plots give values of the BET constant, C,

TABLE 25.2
Some of the proposed values of σ_m for krypton at 77.5.K

Method of calibration	σ_m/nm^2 $molec^{-1}$
Standard anatase	0.195
	0.218
N_2BET at 77.5K	0.195
σ_m =0.162 nm^2 $molec^{-1}$	0.208
	0.218
	0.226
	0.236
Liquid density	0.152
Solid density	0.140

which are less than 100. Tabel 25.3 shows how the coverage at the two extremes of the BET range, and the coverage at the statistical monolayer point varies with *C*. It can be seen that when C > 200 then the BET range corresponds approximately to $1 < \theta < 1.5$, and that when $C < 5$ to $\theta < 1$. However, in the range of C values frequently found for krypton adsorption, $10 < C < 100$, the statistical coverage will vary from first layer to half-filled second layer. In the light of this it is worth considering the relative molecular packing and the adsorption forces in the first and second layers. The molecular packing may well be different in the first and second layers, and therefore the value of σ_m used in BET calculations will be an averge of some sort that reflects this.

Singleton and Halsey and Bonnetain *et al* have argued that because of localisation or psudo-localisation of the molecules in the first layer, the quasi-lattice so formed influences the stacking of molecules in the second layer and that the multilayer steps such as those found by Thomy and Duval, are only found when the adsorbate and adsorbent crystal lattices are geometrically compatible. On this basis Gregg and Sing argue that the application of the BET procedure to any gas at temperatures below its triple point is questionable. However, problems associated with localisation, or perhaps what is better called hindered transalation, are much more general than that, and krypton merely represents another awkward exmple.

TABLE 25.3

The values of the fractional coverage, θ, at relative pressure values $x = 0.05$ and $x = 0.35$, and the value of x when $\theta = 1$, as a function of the BET constant, C.

C	θ at $x = 0.05$	θ at $x = 0.35$	x when $\theta = 1$
2	0.100	0.798	0.415
5	0.219	1.122	0.309
10	0.363	1.297	0.241
30	0.644	1.449	0.155
60	0.799	1.492	0.115
100	0.855	1.510	0.091
200	0.961	1.524	0.066
500	1.014	1.533	0.043

The spacing of the adsorbed atoms in a localised film will be entirely determined by the adsorption potential well distribution, which is in turn determined by the surface structure of the adsorbate. This must of necessity vary from substance to substance and will also be affected in some measure by surface contamination and surface defect concentration. In this case a universal figure for the molecular adsorption cross-section is clearly unattainable, but it may be possible to assign unambiguously a value for a particular adsorbent, although this has not been achieved so far.

For a truly mobile film it is probably reasonable to assume, particularly in the light of the other assumptions in the BET theory, that the packing may approximate to that in the corresponding bulk phase, be it liquid or solid. Emmett and Brunauer showed the general relationship would be

$$\sigma_m = f\left[\frac{M}{\rho L}\right]^{2/3} \times 10^{14}\ \text{nm}^2\ \text{molec}^{-1} \quad (8)$$

where M is the molecular weight, ρ the density of the appropriate phase and L Avogadro's Number. The packing factor, f, depends on the geometry of the adsorbed phase and equals 1.091, for hexagonal close packing. Calculations based on the liquid and solid densities yield values which must represent lower limits for σ_m, since one cannot envisage atoms closer packed than in the solid state.

That some degree of localisation is apparently found with the majority of gases can be seen in Table 25.4 where experimentally required values of σ_m exceed those predicted by equation (8) except when the likely restriction of rotation in the adsorbed film of a non-spherically symmetric molecule, can cause an increase in packing density. In this table the assumption is being made that an adsorbed N_2 film is totally mobile.

TABLE 25.4
Experimental values of σ_m for various gases compared with values derived from the liquid density.

Gas	*Temperature/K*	*σ_m/nm^2 molec^{-1}* Eqn. 8	*Exptl.*
N_2	77.5	0.162	0.162
Ar	77.5	0.128	0.169
Kr	77.5	0.152	0.195
CO	77.5	0.160	0.147
C_5H_{12}	293	0.362	0.523
C_6H_{14}	273	0.390	0.589
C_7H_{16}	298	0.440	0.640

26

Determination of Surface Heterogeneity

Introduction

In the methods about to be described, an attempt is made to use the monolayer models discussed in the previous chapter and an experimentally determined isotherm, to deduce a distribution of adsorption potentials. Such a distribution will be in part dependent on the nature of the substrate surface, but there will also be effect due to the nature of the adsorbate gas. There are in essence two approaches in the literature:

(i) To calculate isotherms from postulated distributions and then compare these with the experimental isotherm. In practice this is limited by the scope of the prior determined comparison isotherms.

(ii) To make use of the calculating power of modern computers to speed up an approach somewhat similar to that of Adamson, in which a distribution is deduced by a successive approximation technique from the experimental isotherm. This has been done by Ross and Morrison and by House and Jaycock. The method of the latter authors will be used as an example.

The algorithm HILDA (Heterogeneity Investigated by a Loughborough Distribution Analysis) was developed to determine the adsorption energy distribution function from the integral equation for the adsorption isotherm on a 'patchwise' heterogeneous surface:

$$V^s = V^s{}_m \int_0^\infty F'(U)\theta(P,U)dU \tag{1}$$

where V^s is the amount of gas adsorbed at the equilibrium pressure P; $F(U)$ is a numerical function of the fraction of the surface with energies U or less, hence $dF(U)/dU = F'(U)$, is the frequency of 'patches' per unit energy interval; θ (P,U) is the fractional coverage of a unisorptic patch given by the model isotherm function; U is equal the the maximum adsorption energy, ${}_mU_0$, minus the zero state vibrational energy, and is taken as positive in magnitude; $V^s{}_m$ may be considered in this case as a normalising factor such that

$$\int_0^\infty F'(U)dU = 1 \tag{2}$$

A direct application of equation (1) is not practicable since some evaluation of the limits of U must be obtained. The method adopted in HILDA was originally suggested by Berenyi. If the temperature and adsorbate-adsorbent interactions are such that the Hill-de Boer or Fowler-Guggenheim model isotherm is subcritical, a phase transition will be located at (P/K), and hence given the lowest pressure P_1 of the data set P, the corresponding K_1 value (related to the adsorption energy) may be obtained on the energy 'patch' where the adsorbate will be at the phase transition. This K_l is then used to determine the upper limit to the adsorption energy distribution that will be computed, U_h. 'Patches' of energy > U_h are assumed to be full. This approximation avoids the difficulty of proposing a distribution of adsorption energies for U in the range U_h to ∞. Another method would be to determine the amount of adsorption at a particular pressure through equation 1 by assuming that there are no 'patches' of energy > U_h. As expected, this leads to an increase in F' at U_h in the final distribution. Obviously information concerning F' is lost near U_h if the isotherm data are not extensive enough; in general, a $V^{s\,l}/V^s{}_N < 0.05$, where N is the total number of data points, is found to be satisfactory. The lower limit, U_l may be obtained from the highest pressure, P_N.

"Patches' of energy < U_l are assumed to be empty. If the model isotherm is supercritical the U_h and U_l limits are obtained by determining $(P/K)_t$ from the point of inflection of the Hill-de Boer and Fowler-Guggenheim model isotherm functions or the value at $\theta = 0.5$ for the Langmuir isotherm. The values of $(P/K)_t$ are available for the Fowler-Guggenheim and Hill-de Boer functions.

Using an initial approximation to (F(U), the problem is to reduce the root mean square (rms) deviation from the experimental data to a minimum.

$$\text{rms} = N^{-1}\left\{\sum_{i=1}^{N}\left[\frac{(V_i^s/V^s_m) - \int_{\theta_l(P_i)}^{\theta_h(P_i)} F(U)d\theta + F_1\,[1 - \theta_h(P_i)] + F_N\theta_l(P_i)}{V^s_i/V^s_m}\right]^2\right\}^{1/2} \tag{3}$$

The iterative improvement is obtained by adjusting $F(U)$:

$$F^c_i = F_i^{c-1}\left\{\frac{V^s_i/V^s_m}{\int_{\theta_l(P_i)}^{\theta_h(P_i)} F^{c-1}(U)d\theta + F_1^{c-1}\,[1 - \theta_h(P_i)] + F_N^{c-1}\theta_l(P_i)}\right\} \tag{4}$$

with the condition that if

$$F^c_i < F^c_{i-1} \tag{5}$$

then

$$F^c_i = F^c_{i-1} + 10^{-8} \tag{6}$$

and normalising:

$$V^s{}_m{}^c = V^s{}_m{}^{c-1} \times F_N \tag{7}$$

$$F^c_i = F_i^{c-1}/F_N \tag{8}$$

where the superscript c labels the cycle. The values of θ_h and θ_l are determined from the isotherm function using the limiting values of *P*. In this method all the F^c_i, values are changed after each literation to obtain an new set of F_i^{c+1} values, with the V^s_m from equations (7) and (8) until the rms reaches a minimum.

For a rapid convergence using this procedure the choice of the initial F *(U)* is important. The method of Adamson and Ling chooses the amount adsorbed at monolayer coverage from a BET analysis, V^s_m (BET), and determines the first approximation to *F(U)* through

$$F^1_i = V^s_i / V^s_m \text{(BET)} \tag{9}$$

This method suffers from the disadvantage that V_m (BET) is not always known or reliable and that the subsequent distribution function, *F'*, is not normalised, that is, $F_n \neq 1$. The normalising procedure (equations (7) and (8) used here allows an arbitrary choice of V^s_m which is adjusted in subsequent cycles to give $F_N = 1$. The initial approximation adopted here is

$$F'_i = V^s_i / V^s_N \tag{10}$$

$$V^s{}_m{}^1 = V^s_N \tag{11}$$

The algorithm used in HILDA represents the *F(U)* and *F(K)* points by cubic splines. Facilities are available for smoothing the data and making multilayer adsorption corrections. The integrals shown in equations (3) and (4) are evaluated using the numerical method of Clenshaw and Curtis. The final distribution function is determined in the range of U_l to U_h. The function *F'* is generated either by using the cubic spline fits to *F(U)* or by determining the gradient by a linear interpolation between consecutive points.

The results of this type of analysis for the isotherm data is shown in Fig. 26.1, in which the Hill-de Boer equation was used as the isotherm function. This illustrates the growth of the main peak on annealing, which corresponds to the development of the (100) face of NaCl, and the decrease in the size of subsidary peaks which are likely to correspond to higher order crystal planes.

Although the distribution obtained using the Fowler-Guggenheim equation for the isotherm function differs slightly, it is still of the same general shape, and has the major peak in virtually the same position on the energy scale. The isotherms

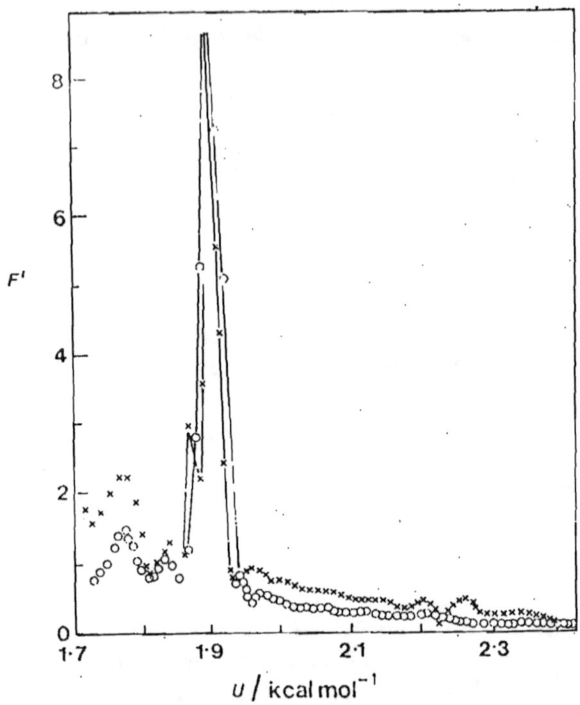

Fig. 26.1: Site energy distribution function obtained by HILDA from the analysis of the data using the Hill-de Boer isotherm function. x unannealed sample. o, annealed sample.

for argon adsorption on the same samples gives a similar picture of what the process involved in annealing sodium chloride might be (see Fig. 26.2

The chief limitations of the method at present appears to be the detailed experimental isotherms required, and the accuracy required. The methods are still undergoing development and refinement. It should be remembered that an apparently good fit obtained with a particular isotherm function and resulting distribution function does not prove the correctness of both. It only indicates that the experimental isotherm can be accurately fitted by such a model. In view of the comments made earlier about hindered translation, and the fact that the model isotherm functions represent extremes of film mobility, no more should be envisaged.

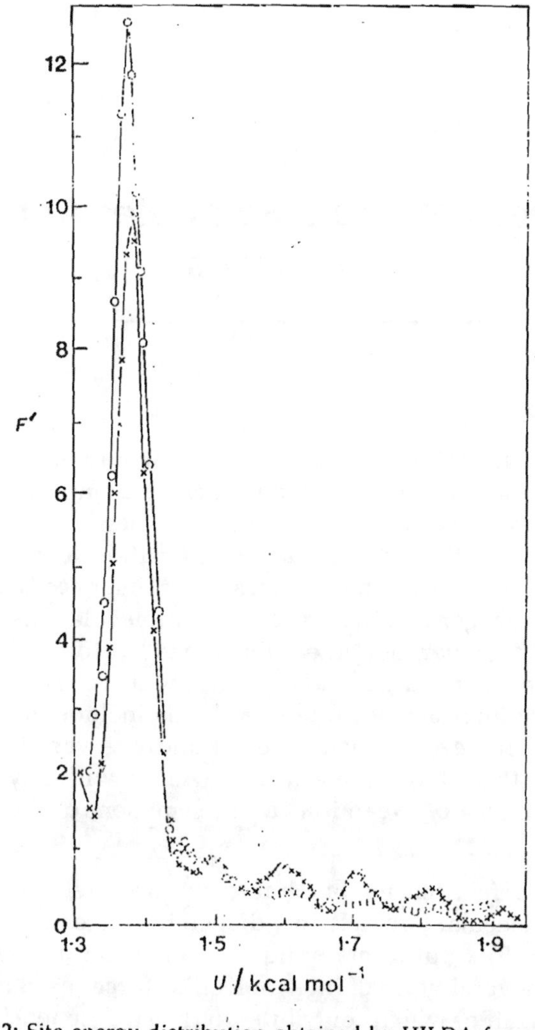

Fig. 26.2: Site energy distribution obtained by HILDA from the analysis of argon adsorption data on the same NaCl samples as used for Fig. 26.1. X unannealed sample. o, anneaaled sample.

27

Forces Between Molecules Across Interfaces

Introduction

From what has already been said it would be reasonable to assume that the region of change across an interface would only be a few molecular diameters in thickness, beyond which the behaviour reflects that of the bulk material. However, it has been suggested that long range forces exist at interfaces, extending over much larger distances (~10 -100 molecular diameters), although the experimental evidence has led to much debate over the years. In our general consideration of the interactions that exist at interfaces we often need only include the London dispersion force (often called the London - van der Waals force) and the retardation effects, together with any electric forces that exist between ions and permanent dipoles in the system.

The surface tension of a liquid γ^{LV} is a direct measure of the intermolecular forces. If only dispersion forces are involved them $\gamma^{LV} = {}^{d}\gamma^{L}$. But in many liquids this is not the case. For example, several types of intermolecular forces exist in liquid water (W), all of which contribute to the surface tension γ^{WV}. Fowkes proposed that we write

$$\gamma^{WV} = {}^{d}\gamma^{W} + {}^{h}\gamma^{w} \tag{1}$$

where ${}^{d}\gamma^{w}$ is the contribution from dispersion forces, and ${}^{h}\gamma^{W}$ that due to hydrogen bonding and dipole interactions. For mercury (M) there is a contribution due to metallic bonds, hence

$$\gamma^{MV} = {}^{d}\gamma^{M} + {}^{m}\gamma^{M} \quad (2)$$

The interface between two immiscible liquids, for example water and oil (O), consists essentially of adjacent monolayers of the two liquids, and the interfacial tension is the sum of the two individual tensions. Both liquids have, when separated, their normal surface tension, but when in contact the molecules in the water monolayer are subject to forces from both the bulk water and the oil molecules, hence the tension is diminished — similarly for the oil monolayer. The magnitude of the oil-water force is a function of the dispersion forces of the oil and water molecules; it has been shown that the geometric mean of the contributions to surface tension of these forces $({}^{d}\gamma^{O}\,{}^{d}\gamma^{w})^{1/2}$, accurately describes the force of interaction. The tension in the oil layer is reduced by the same quantity, hence

$$\gamma^{ow} = \gamma^{ov} + \gamma^{WV} - 2\sqrt{({}^{d}\gamma^{o}\,{}^{d}\gamma^{w})} \quad (3)$$

where γ^{ov} and γ^{wv} are the surface tensions of oil and water respectively.

In general for any two immiscible liquids 1 and 2, an estimate for ${}^{d}\gamma^{1}$ (or ${}^{d}\gamma^{2}$) may be obtained by summing the pair potentials of all the surface volume elements with all the volume elements below the surface, giving

$$\begin{aligned} dH &= dU + PdV + VdP - \gamma dA - Ad\gamma \\ &= TdS + VdP - Ad\gamma + \Sigma_i \mu_i dn_i \quad (4) \\ dA &= dU - TdS - SdT \\ &= - SdT - PdV + \gamma dA + \Sigma_i \mu_i dn_i \end{aligned}$$

which is identical with the equation deduced by classical methods. However, the analogus expression to equation is different:

$$\begin{aligned} dG &= dH - TdS - SdT \\ &= - SdT + VdP - Ad\gamma + \Sigma_i \mu_i dn_i \quad (5) \end{aligned}$$

which equations give rise to the expressions for the chemical potential

$$\mu_i = \left[\frac{\partial U}{\partial n_i}\right]_{nj,\ S,V,A}$$

$$= \left[\frac{\partial H}{\partial n_i}\right]_{n_j, S, P, \gamma}$$

$$= \left[\frac{\partial A}{\partial n_i}\right]_{n_j, T, V, A}$$

$$= \left[\frac{\partial G}{\partial n_i}\right]_{n_j, T, P, \gamma} \tag{6}$$

all of which would appear consistent with the results for a bulk phase alone. By integration of the above equation expressions for the functions themselves may be obtained. However, since the expressions for dU and dA are identical in both cases the integrated expressions must be identical, but the results for the remaining functions are

$$H = TS + \Sigma_i \mu_i n_i \tag{7}$$

and

$$G = \Sigma_i \mu_i n_i \tag{8}$$

This latter result is important since it is of the form familiar from the treatment of bulk phases, and hence is of universal applicability, whereas the function G gives rise to the peculiar equation for a surface phase.

The methods of Guggenheim and Hill, although involving re-definition of H and G, lead to a more consistent treatment of interphases than what may be termed classical methods. Using their approach the significance of thermodynamic parameters is not longer altered by the presence or absence of an interphase in the system. In fact the definitions of H and G are precise generalised definitions, and H and G should be regarded as specific cases for bulk phases in the absence of an interphase.

GIBBS -Duhem Equation for a Interphase

We have previously deduced that

$$dG = -SdT + VdP - Ad\gamma + \Sigma_i \mu_i dn_i \quad (9)$$

and if equation is differentiated

$$dG = \Sigma_i \mu_i dn_i + \Sigma_i n_i du_i \quad (10)$$

Thus by subtracting these two equations

$$SdT - VdP + Ad\gamma + \Sigma_i n_i d\mu_i = 0 \quad (11)$$

which is the analogue for an interphase of the usual form of the Gibbs-Duhem equation for a bulk phase which merely omits the term $Ad\gamma$.

28

Adsorption from Solution at Solid-Liquid Interface

Introduction

We shall now consider the interface between a solid and a liquid containing two components. At equilibrium there is a distribution of each of the components between the bulk liquid and the interphase, and in general terms this is determined by the relative strengths of the interactions of the solid surface with each component and of those between the components in the liquid phase. The composition of the interphase is a function of the concentration of solution and the temperature. There is a wealth of information on the adsorption of components from solution at the solid-liquid interface, usually expressed as adsorption isotherms; but rarely at more than one temperature, and even fewer measurements have been made of the heat effects by immersional calorimetry, hence thermodynamic data are scarce.

It is convenient to divide this subject into two parts. The first relates to solutions for which there is complete miscibility of the two components over the whole concentration range, as in binary liquid mixutres. The second is concerned with solutions that are sufficiently dilute for there to be no significant change in the activity of the solvent (the component in excess) during adsorption. This is the most common situation in practice, such as with dilute solutions of electrolytes, polymers and surface active agents which arise in many technological and natural systems. The second is essentially a special case of the first, as the following simple algebraic analysis demonstrates:

Taking m g of a solid in contact with a solution which initially contained n_1^o moles of component 1 at mole fraction x_1^o, and n_2^o moles of 2 at x^o_2, and after adsorption equilibrium is reached the solution contains n_1 moles at x_1, and n_2 at x_2, then the change on adsorption is given by

$$
\begin{aligned}
n_1^s = m = n_1^o &- n_1 \\
&= n^o x_1^o - x_1(n_1 + n_2) \\
&= n^o x_1^o - x_1(n^o - n_1^s m - n_2^s m) \qquad (1)
\end{aligned}
$$

or

$$n^o(x_1^o - x_1)/m = n_1^s x_2 - n_2^s x_1 \qquad (2)$$

where n_1^s and n_2^s moles of the components are adsorbed per gram of solid and n^o ($= n_1^o + n_2^o$) is the total number of moles of liquid components present in the system. In equation (2) we have the measurable quantities on the left and the unknowns on the right, and a plot of n^o $(x^o_1 - x_1)/m$ versus x_1 is an adsorption isotherm but with the proviso that it is not an isotherm of actual moles of component 1 adsorbed - examination of the limiting values of the adsorption function at $x_1 = 0$ and 1 serves to illustrate the point. The measured quantity is an 'apparent' adsorption and represents the change in this quantity as the relative concentrations of the two componets are varied across the whole range of mole fraction. The isotherm, being a function of the behaviour of two solution components, is commonly, called a *composite* isotherm. Mathematical analysis, with assumptions, permits evaluation of the *individual* isotherms for the two components, as discussed later.

In the case of dilute solutions, for which $x_2 \approx 1$ and $x_1 \rightarrow 0$, the measured change in solution concentration on adsorption gives directly the individual isotherm (n_1^s) for the solute'.

Adsorption from binary liquid mixutres

The apparent adsorption is simply related to the surface excess functions which arise from the Gibbs adsorption equation with the appropriate choice of dividing surface. Three cases are usually considered:

(i) with the dividing surface located at the surface of the solid (the solid is assumed to be insoluble), the excess function being $\Gamma^s{}_1$;

(ii) with the dividing surface defined such that the excess of the other component is zero, giving $^2\Gamma_1$;

(iii) taking the excess of component 1 in one m^2 cross-section of surface region over the moles that would be present in a region of bulk solution containing the same total number of moles as the surface region considered, giving $^n\Gamma_1$. The excess functions are related to each other and to the apparent adsorption (cf equation (2)) by the following

$$n^o \Delta x_1/ms = \Gamma^s{}_1 x_2 - \Gamma^s{}_2 x_1 = {}^n\Gamma_1 = {}^2\Gamma_1 x_2 \tag{3}$$

where s is the specific surface area of the adsorbent. The reader is recommended to derive these relationships, since this helps considerably in the understanding of the principle of excess quantities.

Composite adsorption isotherms have a variety of shapes depending on the adsorption behaviour of the individual components. Ostwald and de Izaguirre were the first to point out, in 1922, that by postulating the shape of the individual isotherms the use of simple mathematical analysis led to Composite isotherms with shapes that are now well documented (Fig. 28.1). Considering the entire range of mole fraction it is necessary for the apparent adsorption of component 1 to go through a maximum, even if over a large part of the concentration range there is no adsorption of component 2. The apparent adsorption of component 1 can also take negative values but only when there is adsorption of the other component. So composite isotherms for binary liquid mixutres exhibit one maximum (of necessity), and frequently one minimum. These principles apply to adsorption at both liquid-vapour and solid-liquid interfaces, and there is merit in comparing the behaviour of the two situations for any one liquid system; this is conveniently done in terms of the excess functions $^n\Gamma_1$ and $^n\Gamma_2$. From such a comparison it should be possible to examine the specific effect of the solid on the adsorption process, if one exists; for example for an oxide in contact with an alcohol + hydrocarbon mixutre we might expect the adsorption to be affected in a significant way by the alcohol-oxide interactions, in contrast to the behaviour at the solution-vapour interface and this should be reflected in a comparison of the composite isotherms for the two situations.

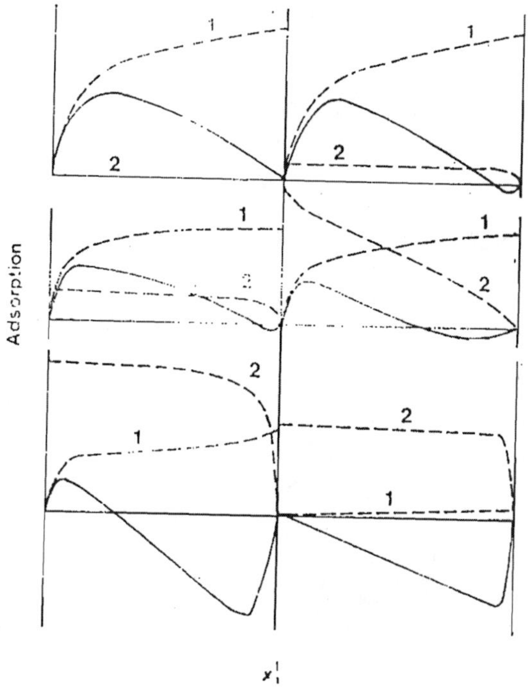

Fig. 28.1: Shapes of composite isotherms (continuous lines) derived by simple mathematical analysis from postulated individual isotherms for components 1 'solute', and 2 'solvent'.

Schay has recognised five types of composite isotherms for adsorption at the solid-liquid interface, as illustrated in Fig. 28.2. These five types are also found for adsorption at the liquid-vapour interface, and an examination of the thermodynamics of this system is of value in the interpretation of isotherm shape.

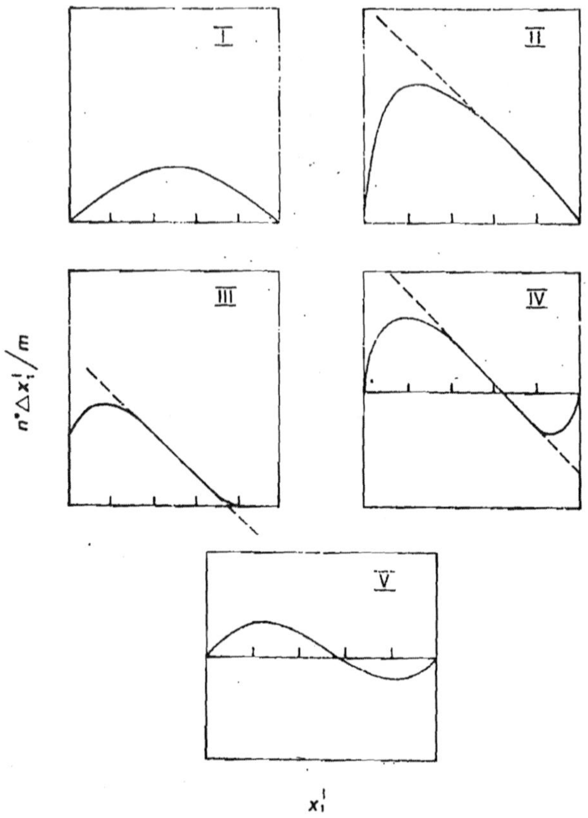

Fig. 28.2: Schay's five types of composite isotherms.

For component i in solution at equilibrium the chemical potential in the surface phase (liquid-vapour interface) is

$$\mu_i^s = \mu_i^l + A_i \gamma^l \tag{4}$$

where superscript l refers to bulk solution, A_i is the molar area of the component in the surface and γ^l is the surface tension of the solution. Taking the pure liquid as the standard state

$$\mu_i^s = \mu_i^{\varnothing s} + RT \ln a_i^s \tag{5}$$

$$\mu_i^l = \mu_i^{\varnothing l} + RT \ln a_i^l \tag{6}$$

where a_i is the activity of the component. Hence at equilibrium

$$\mu_i^{\varnothing s} + RT \ln a_i^s = \mu_i^{\varnothing l} + RT \ln a_i^l + A_i \gamma^l \tag{7}$$

Since

$$\mu_1^{\varnothing s} - \mu_1^{\varnothing l} = A_i \gamma_i^l \tag{8}$$

then for component 1

$$RT \ln \frac{a_1^l}{a_1^s} = A_1(\gamma_1^{LV} - \gamma^l) \tag{9}$$

and similarly for component 2. Hence

$$\left(\frac{a_1^s}{a_1^l}\right)^{1/A_1} \left(\frac{a_2^l}{a_2^s}\right)^{1/A_2} = K = \exp\left[(\gamma_2^{LV} - \gamma_1^{LV})/RT\right] \tag{10}$$

or for $A_1 = A_2 = A$

$$K = \frac{a_1^s a_2^l}{a_1^l a_2^s} = \exp\left[\frac{(\gamma_2^{LV} - \gamma_1^{LV}) A}{RT}\right] \tag{11}$$

where K is the themodynamic equilibrium constant for the exchange (adsorption) process as represented by the expression

$$\frac{1}{A_1}(1)^l + \frac{1}{A_2}(2)^s = \frac{1}{A_1}(1)^s + \frac{1}{A_2}(2)^l \tag{12}$$

From equation (11) it is not immediately apparent which factors predominate in deciding the shape of the composite isotherm. Interpretation of adsorption data requires accurate values of surface tension and activity coefficients, and to have such comprehensive information on any one system is not common. Some guiding principles are established. Clearly the large the difference in surface tensions of the two pure components the larger is K, and the larger the adsorption of one component. Type I adsorption in Schay's classification would, at the liquid-vapour interface, be associated with mixtures of liquids that are similar in type, and exhibit only small deviations from ideality, for example toluene/benzene with $\gamma_2^{LV} - \gamma_1^{LV} < 1$ mN m^{-1}. For non-ideal systems with a fairly large ($\gamma_2^{LV} - \gamma_1^{LV}$), type II adsorption is obtained, for example ethanol ($\gamma^{LV} = 22.0$ mN m^{-1}) and benzene ($\gamma^{LV} = 28.2$ mN m^{-1}) mixture. Larger values of ($\gamma_2^{LV} - \gamma_1^{LV}$) and large departures from ideality lead to type III adsorption, and with smaller surface tension differences but with considerable non-ideality so the S-shaped curves of types IV and V adsorption result. Examples are: type III, acetone/water ($\gamma_2^{LV} - \gamma_1^{LV} = 49.69$ mNm^{-1}); type IV, methanol/ benzene ($\gamma_2^{LV} - \gamma_1^{LV} = 6.1$ mNm^{-1}); type V, ethanoic acid/ benzene ($\gamma_2^{LV} - \gamma_1^{LV} = 1.32$ mNm^{-1}).

The interest lies in a comparison of the liquid-vapour isotherms with those for the solid-liquid interface - any differences might be interpreted in terms of specific effect of the solid surface on the composition and character of the adsorbed layer. Normally at the liquid-vapour interface it is the component with the lower surface tension that is adsorbed preferentially over the whole range of mole fraction (Types I, II and III), but where a maximum or minimum occurs in the surface tension curve an S-shaped composite isotherm is obtained (type IV and V) corresponding to an increased contribution of the other components; and it may be preferentially adsorbed over the whole range in the extreme case. At the solid-liquid interface preference may be for either component depending, it would seem, on the chemical nature of adsorbate and adsorbent, hence on the nature of the forces existing between them. Simple comparisons are made in Fig. 28.3 for chloroform + benzene mixtures with and without boehmite, and for cyclohexane + benzene mixtures at the liquid-vapour interface and on the carbon black Spheron 6, charocoal and boehmite. In the first case the component with the lower surface tension (chloroform) is preferentially adsorbed at both interfaces, but the adsorption is larger when the solid is present, presumably because of the enhanced interaction between the chloroform dipole and the oxide surface. In the second case that with the lower tension (cyclohexane) is preferentially adsorbed at the liquid-vapour interface, but vice versa at the solid liquid interface reflecting the force between benzene and carbon (through π bonds)

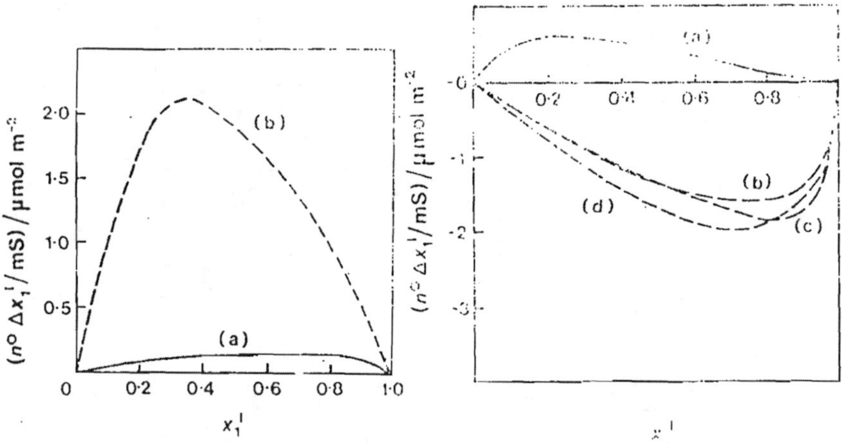

Fig. 28.3: Comparison of composite isotherms for adsorption at the solid-liquid and liquid-vapour interfaces. Left - adsorption from mixtures of benzene and chloroform at the liquid-vapour interface (a), and on boehmite (b). Right - adsorption from mixtures of benzene and cyclohexane at the liquid-vapour interface (a), on Spheron 6 (b), on charcoal (c) and on boehmite (d).

which, in a system where all the forces are relatively weak, easily tips the balance in favour of the higher surface tension component being preferentially adsorbed. Fig. 28.4 shown the simple case of 1,4-dimethylbenzene + heptane at the liquid-vapour interface, and at rutile-solution interfaces with the surface forces of the solid being modified by pre-adsorbed water. With the water-free surface the S-shaped isotherm arises from the fact that the two solution components interact in a different manner with different regions of the surface (on a molecular scale), which disappears when the surface is covered with water and on which the armatic molecule has preference. Few such analyses have been carried out since appropriate data are scarce.

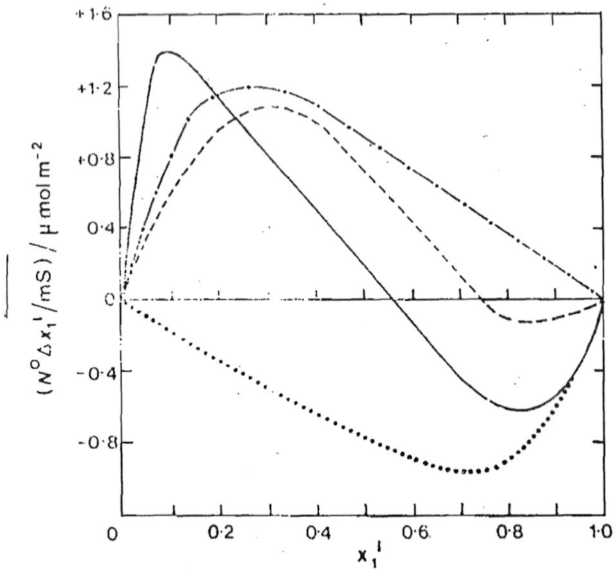

Fig. 28.4: Composite isotherms for adsorption from 1,4-dimethylbenzene/ heptane mixtures on rutile ritanium (IV) oxide, free of adsorbed water (——), wit approximately I monolayer of adsorbed water (- - - - -), with 2-3 monolayers of adsorbed water (-.-.), compared with that for the liquid-vapour interface (. . .).

Everett's thermodynamic analysis of adsorption from binary liquid mixutres demonstrates, using a lattice model and $A_1 = A_2$, that the perfect system (all activity coefficients equal to unity) gives types I and II composite isotherms, with the trend from I to II increasing with *K*. S-Shaped isotherms are predicted by applying regular solution theory, although azeotropic behaviour is only evident when *K is* close to unity, that is, $\Upsilon_1^{LV} \approx \gamma_2^{LV}$, in general agreement with experiment (see above). So for non-ideal liquid mixutres the shape of the isotherm is determined by (a) the difference in adsorption energies of the two components and (b) the nature of the bulk and adsorbed phases as reflected in the activity coefficients — if (a) is large we expect types I and II, if small types IV and V with the relative proportions of positive and negative regions determined by the departure from ideality in the bulk and/or adsorbed phases. U-shaped isotherms are also predicted for molecules of different sizes assuming athermal solution behaviour, the preferential adsorption being enhanced by disparity of molecular size; for non-athermal solutions S-shaped isotherms are predicted.

In the majority of cases the composite adsorption behaviour is only of academic interest, since it gives no direct information on the actual composition of the adsorbed phase which is often required in practical application of solution adsorption. For want of anything better it has been practice to assume that the adsorption is confined to a monomolecular layer. This is acceptable to a good approximation if (a) the two molecules involved occupy approximately equal areas at the surface and (b) the adsorbate and adsorbent are such that interaction between them extends only over one molecular distance (as discussed earlier). On this basis we may take

$$n_1^s/(n_i^s)_m + n_2^s(n_2^s)_m = 1 \qquad (13)$$

where $(n^s_1)_m$ and $(n_2^s)_m$ are the values of the surface concentrations when the surface is covered with a monolayer of the single component. Combining equations (13) and (2) provides sufficient data from the calculation of the individual isotherms. Monolayer values are derived from vapour adsorption data or molecular models, and although in principle they may be inaccurate for the use to which they are applied, since for the mixed monolayer the packing and orientation may differ from those in a single component monolayer, there has been a large number of such calculations made with varied success.

The analysis is fraught with difficulties. Take the case of a mixture of 1-octanol and 1,4-dimethylbenzene in contact with a rutile titanium (IV) oxide surface. A reasonable value of $(n^s_I)_m$ for the alcohol is equivalent to 0.2 nm^2 molec-1, the cross-sectional area of a hydrocarbon chain close packed on a water surface (Langmuir-Adam trough studies) — this assumes that all concentrations the alcohol adopts a vertical orientation. For 1,4-dimethylbenzene we take 0.45 nm^2 $molec^{-1}$ which is obtained from a molecular model. Using these monolayer values with composite adsorption data the individual isotherms were theoretically unacceptable since they exhibited maxima and minima. But for this system the chain length of the adsorbed alcohol is significantly larger than the thickness of the hydrocarbon adsorbed in parallel orientation. Hence there is effectively more than one layer of 1,4-dimethylbenzene molecules involved in the monolayer of thickness corresponding to the length of the octanol molecule. Allowing for this effect leads to acceptable individual isothrms showing alcohol adsorption increasing with concentration in a manner (Langmuirian) that 'experience' would predict; there is no other way to predict isotherm shape. So we have 'constructed' a simple multilayer.

Everett has also considered a multilayer model for monomer-dimer mixtures involving both parallel and perpendicualar orientations, and shows that much larger surface excesses are to be expected than for the case when all molecules are in parallel orientation as a monomolecular layer, when there are strong positive deviations from ideality in solution, that is, the more non-ideal the greater the contribution from the second and higher layers. Multilayer theory also predicts S-shaped composite isotherms. The predominant orientations is seen to be dependent on the relative strenghts of the interaction of the two segments of the dimer with the surface — if only one is preferentially adsorbed (as expected in the octanol/dimethylbenzene/rutile case considered earlier) the orientation is expected to change from parallel to perpendicular as the solution concentration is increased, although the predominant orientation is perpendicular. The concept of changing orientation with concentration makes use of equation (13) pointless.

To some extent thoery leads practice in this field, but the experimental data required are now defined and only a few presently available satisfy all the criteria of extent and accuracy.

29

Gibbs Adsorption Isotherm

Introduction

The quantity of the components adsorbed at the interface is obviously a significant parameter, and particularly of interest is the relationship between the extent of adsorption and the interfacial tension. Since it is difficult to make systematic measurement when many factors vary simultaneously, it is convenient to investigate the relationship between the extent of adsorption and interfacial tension whilst the temperature is held constant; hence the adsorption isothem.

The original derivation of what is variously known as either the Gibbs adsorption isotherm or the Gibbs equation is somewhat more difficult to follow than that starting from the Gibbs-Duhem equation.

The Gibbs-Duhem relation for an interphase under conditions of constant temperature and pressure gives

$$-A d\gamma = \Sigma_i n_i d\mu_i \tag{1}$$

If we now write

$$\Gamma_i = \frac{n_i}{A} \tag{2}$$

where Γ_i represents the amount of species *i* in unit area of the interphase, then we have the Gibbs adsorption isotherm

$$-d\gamma = \Sigma_i \Gamma_i d\mu_i \tag{3}$$

where the quantity Γ_i requires exact definition which is dependent on the choice of boundary or dividing surface for the interphase.

Choice of a Dividing Surface

It is easiest to illustrate the use of such an equation by references to a two-component liquid mixture, and the example frequently chosen to illustrate the point is that of ethanol and water, where water is the solvent and ethanol the solute.

The Gibbs adsorption isotherm for a two-component system may be written as

$$—d\gamma = \Gamma_1 \, d\mu_1 + \Gamma_2 d\mu_2 \qquad (4)$$

where the subscript 1 refers to the solvent and subscript 2 to the solute. Whilst recognising that the interfacial region is best regarded as an interphase. This dividing surface may be considered to be positioned so as to give rise to simplification of equation (4). Gibbs defined the position of the dividing surface such that the surface excess of solvent is zero, and hence we may write

$$—d\gamma = {}^1\Gamma_2 \, d\mu_2 \qquad (5)$$

where ${}^1\Gamma_2$ is now termed the surface excess of the solute with the dividing surface so defined. The significance of this may be seen from the Fig. 29.1. To quote from Gibbs' original paper:

> It is generally possible to place the dividing surface so that the total quantity of any desired component in the vicinity of the surface of discontinuity shall be the same as if the density of that component were uniform on each side quite up to the dividing surface. In other words, we may place the dividing surface so as make any one of the quantities Γ_1, Γ_2, etc., vanish.

Thus the dividing surface in Fig. 29.1 is such that the surface excess of solvent is zero, that is the shaded area ABE will be equal to the shaded area EGH. The difference between areas FGI and CDF will give *the surface excess of solute* ${}^1\Gamma_2$, *which is defined as the algebraic excess of solute in unit cross-section of the surface region over that which would be present in a bulk region containing the same number of moles of solvent as does the section of the surface region.*

A corresponding symmetrical definition of ${}^{2}\Gamma_1$ can be written where Γ_2 is equal to zero, and ${}^{2}\Gamma_1$ represents the excess of solvent in unit cross-section of the surface region over that which would be present in a bulk region containing the same number of moles of solute as does the section of the surface region.

The above is not the only way of locating the dividing surface, and several alternative definitions are possible. The dividing surface could be placed such that the algebraic sum areas in Fig. 29.1 to the right of the dividing surface is equal to the sum of the areas to the left of the dividing surface. The surface excess defined in this manner may be written ${}^{n}\Gamma_i$, which represents the excess of the *i*th component in unit cross-section of the surface region over the moles that would be present in a bulk region containing the same total number of moles as does the section of the surface region.

A quantity ${}^{m}\Gamma_i$ may be defined as the excess of the *i*th component in the region of the surface over the number of the moles that would be present in a bulk region of the same total mass as the surface region.

Instead of using the total mass in the definition the total volume may be used. Thus ${}^{v}\Gamma_i$ is the excess of the *i*th component in the region of the surface over the number of moles that would be present in a bulk region of the same total volume as the surface region.

The exact significance of these various definitions of surface excess is somewhat difficult to visualise, and a numerical example is of some assistance. Consider a bulk of solution of ethanol in water, where the mole fraction of each component is 0.5. In order to make the arithmetic easy to follow we will use numbers which are of the right order of magnitude but not experimentally exact, but they will serve the purpose of illustration well enough. If a sample of approximately 0.1 cm^3 of the surface region is taken containing 1 mmol of water and 1.016 mmol of ethanol from a surface area of 2 m^2, it is probable that some of the bulk solution is included in this sample together with the whole of the surface region. In order to calculate the various surface excesses the following data will also be required,

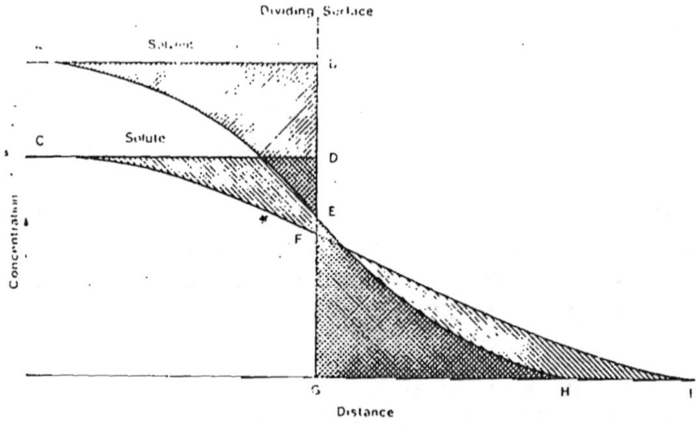

Fig. 29.1: Definition of a dividing surface

namely that the relative molecular weights of water and ethanol are 18 and 46, and that the value of the respective molar volumes are 18 and 58 $cm^3\ mol^{-1}$.

The quantity ${}^1\Gamma_2$ is the excess which 1.016 mmol of ethanol represents over the quantity present in the bulk associated with the same amount of water as in the surface. The surface sample contains 1 mmol of water, and in the bulk 1 mmol of water is associated with 1 mmol of ethanol. Thus the surface excess is equal to $1.016-1 = 0.016$ mmol for an area of 2 m^2. Thus ${}^1\Gamma_2 = 0.016/2\ \text{mmol m}^{-2} = 8\ \mu\text{mol m}^{-2}$.

The symmetrical definition of ${}^2\Gamma_1$ can be calculated as well. If the surface phase sample contains 1.016 mmol of ethanol the surface excess of water will be the excess of 1 mmol of water over the amount of water associated with 1.016 mmol of ethanol in the bulk solution. The surface excess is therefore equal to 1-1.016 = –0.016 mmol for an area of 2 m^2. Thus $^2\Gamma_1$ = –0.016/2 mmol m^{-2} = –8 μmol m^{-2}.

The remaining three definitions are calculated somewhat differently. Firstly we have the quantities $^n\Gamma_1$ and $^n\Gamma_2$, in which the bulk region for comparison contains the same total number of moles as the surface region. If the surface region contains 1 mmol of water and 1.016 mmol of ethanol, then the bulk region for comparison will contain 1.008 mmol water and 1.008 mmol of ethanol. The surface excess of ethanol is 1.016-1.008 = 0.008 mmol and that of water is 1-1.008 = —0.008 mmol for an area of 2 m^2. Thus $^n\Gamma_1$ = -0.008/2 mmol m^{-2} = —4 μmol m^{-2} and $^n\Gamma_2$ = 0.008/2 mmol m^{-2} = 4 μmol m^{-2}.

The next definition relates to the qunatities $^m\Gamma_1$ and $^m\Gamma_2$ in which case the total mass is the basis of comparison. The sample of surface has a total mass of (10^{-3} × 18) + (1.016 × 10^{-3} × 46) = 0.0647 g. This amount of bulk solution contains 18.2 mg of water and 46.5 mg of ethanol, that is 1.011 mmol of water and 1.011 mmol of ethanol. The surface excess of ethanol will be 1.016-1.011 mmol = 0.005 mmol and that of water 1-1.011 = – 0.011 mmol for an area of 2m^2. Thus $^m\Gamma_1$ = -5.5 μmol m^{-2} and $^m\Gamma_2$ = 2.5 μumol m^{-2}.

The remaining definition is on the basis of comparison of equal volumes of surface and bulk regions. The total volume of the surface region is (10^{-3} × 18) + (1.016 × 10^{-3} × 58) = 0.0769 cm^3 and this volume of bulk solution will contain 1.012 mmol of ethanol and of water. The surface excess of ethanol will be 1.016–1.012 = 0.004 mmol and that of water 1-1.012 = – 0.012 mmol for an area of 2 m^2. Thus $^V\Gamma_1$ = –6 μmol m^{-2} and $^vT_2$ = 2 μ mol m^{-2}.

We can regard the term surface excess as a means of expressing surface concentrations in a way that relates such a parameter to the bulk phase composition with which the interphase is in equilibrium. The thermodynamic approach which we have outlined systematises the possible definitions, and the relationship of such quantitites to other basic thermodynamic functions.

Relationship between the variously Defined Surface Excesses

The Gibbs-Duhem relationship for a bulk phase may be written

$$SdT - VdP + \Sigma_i n_i d\mu_i = 0$$

which at constant temperature and pressure reduces to

$$\Sigma_i n_i d\mu_i = 0$$

which may be written for a two-component system as

$$n_1 d\mu_1 + n_2\, d\mu_2 = 0$$

or if mole fractions are employed

$$x_1 d\mu_1 + x_2\, d\mu_2 = 0$$

Substitution in the Gibbs adsorption isotherm as written in equation (4) yields

$$-d\gamma = -\Gamma_1 \frac{x_2}{x_1} d\mu_2 + \Gamma_2 d\mu_2$$

Therefore

$$-\frac{d\gamma}{d\mu_2} = \Gamma_2 - \frac{x_2}{x_1}\Gamma_1 \qquad (6)$$

Since the quantity - $d\gamma/d\mu_2$ is independent of the choice of dividing surface, them it follows that the expression on the right-hand side of equation (6) must also be independent.

It can be seen from the numerical example that the quantities ${}^n\Gamma_i$, ${}^m\Gamma_i$ and ${}^v\Gamma_i$ may be related by equaitons of the type

$$f_1 Q_1 + f_2 Q_2 = 0 \qquad (7)$$

where the factors f_i are unity when Q_i is ${}^n\Gamma_i$, are equal to the molecular weights when Q_i is ${}^m\Gamma_i$, and correspond to the molar volumes when Q_i is ${}^v\Gamma_i$. Combination of equations (6) and (7) gives

$$-\frac{d\gamma}{d\mu_2} = \Gamma_2 \left[1 + \frac{x_2 f_2}{x_1 f_1} \right]$$

and similarly it may be deduced that

$$-\frac{d\gamma}{d\mu_1} = \Gamma_1\left[1 + \frac{x_1 f_1}{x_2 f_2}\right] \tag{9}$$

Therefore the series of relationships may be written

$$-\frac{d\gamma}{d\mu_2} = {}^1\Gamma_2 \tag{10}$$

$$= {}^n\Gamma_2\left[1 + \frac{x_2}{x_1}\right] \tag{11}$$

$$= {}^m\Gamma_2\left[1 + \frac{x_2 M_2}{x_1 M_1}\right] \tag{12}$$

$$= {}^v\Gamma_2\left[1 + \frac{x_2 V_2}{x_1 V_1}\right] \tag{13}$$

where M_i represents the molecular weights and V_i the molar volumes. A corresponding set of relations may be deduced for $-d\gamma/d\mu_1$.

30

Porosity

Introduction

The investigation of many porous solids involves the use of gas adsorption methods. Pores are usually classified according to size in the manner originally proposed by Dubinin. Pores having width less than 2 nm are termed *micropores*, those between 2 and 50 nm, *mesopores*, and those whose width is larger than 50 nm, *macropores*. The first two classifications can be investigated by gas adsorption methods, but the latter appear too similar to a plane surface for adsorption differences to be significant, but they can be investigated by the measurement of the pressure required to force mercury into them, and to convert the volume of mercury forced in at a particular pressure into a pore volume vs. pore radius plot using the Laplace equation.

One approach to adsorption in micropores is by means of potential theory, and was developed by Dubinin *et al.* They assumed that for two different adsorptives filling the micropore adsorption space to the same extent, *W*, the adsorption potentials will possess a constant ratio to one another, thus

$$\frac{\varepsilon_1}{\varepsilon_2} = \beta \qquad (1)$$

where β is termed the affinity coefficient. If adsorbate 2 is chosen as an arbitrary reference standard, then we may write equation (1) as

$$\frac{\varepsilon}{\varepsilon_o} = \beta \qquad (2)$$

where the subscript *o* denotes the standard adsorbate. For this standard we may write

$$W = f(\varepsilon_o) = f(\varepsilon/\beta) . \quad (3)$$

Now if W is a Gaussian function of the adsorption potential, then

$$W = W_o \exp(-\tilde{k}\,\varepsilon^2_o) \quad (4)$$

where $\tilde{k}$ is a constant characterising the pore distribution. Using equations (1) and (2), and also using the fact that $W = n^s/\rho$ where ρ is the molar density of the liquid adsorbate, then

$$\ln n^s = \ln (W_o\rho) - \frac{\tilde{k}\,R^2T^2}{\beta^2} \{\ln (P_o/P)\}^2 \quad (5)$$

which is known as the Dubinin-Radushkevich equation, where W_o is the total micropore volume. Data for an activated carbon made from sucrose for the adsorption of benzene at 293K are shown in Fig. 30.1 plotted according to equation (5)

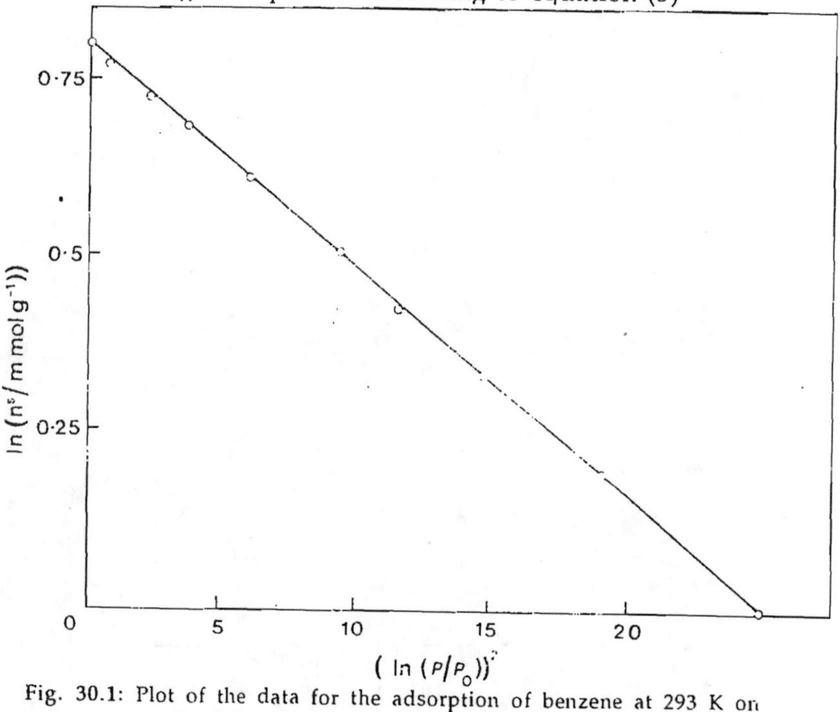

Fig. 30.1: Plot of the data for the adsorption of benzene at 293 K on an activated carbon made from sucrose, using the Dubinin-Radushkevich equation (5).

Kaganer has pointed out that an equation similar in form would be found in the monolayer region for a planar surface Gaussian distribution of adsorption potentials. Thus if we write

$$\theta = \exp(-k_1\varepsilon^2), \tag{6}$$

then equation (4) has the analogous form

$$\ln n^s = \ln n^s_m - k_1 R^2T^2 \{\ln (P_o/P)\}^2 \tag{7}$$

Thus for a microporous solid the value of the micropore volume

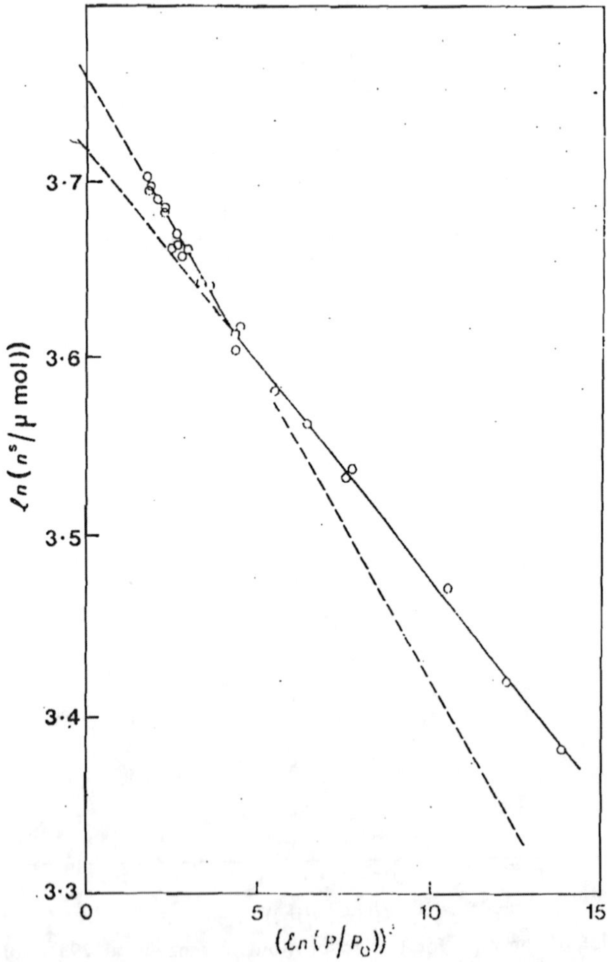

Fig. 30.2: Dubinin-Radushkevich plot for the adsorption of krypton at 77.5 K on tin (IV) oxide.

may be obtained from the intercept on the plot of ln n^s *vs* (ln $(P/P_o))^2$, or the monolayer capacity, n^s_m, from the same plot for a non-porous meaterial, if the assumptions are plausible.

Such plots are not always easy to interpret as can be seen in Fig. 30.2 for the adsorption of krypton on a tin (IV) oxide sample at 77.5K. This sample is thought to be partially microporous, and it is possible that the low pressure region is primarily controlled by adsorption in micropores, and that upturn from this line is due to mesopores or multilayer formation on the external surface of the porous body.

It should be noted that if the adsorbent is entirely microporous a Type I isotherm wil be experimentally determined, and this may sometimes be approximated by Langmuir's equation. It has been shown that an almost linear plot results using Dubinin-Radushkevich axes when $b = 18/P_o$ at pressures where $P/P_o < 0.25$, and that the graphs are appreciably curved at both higher and lower b values.

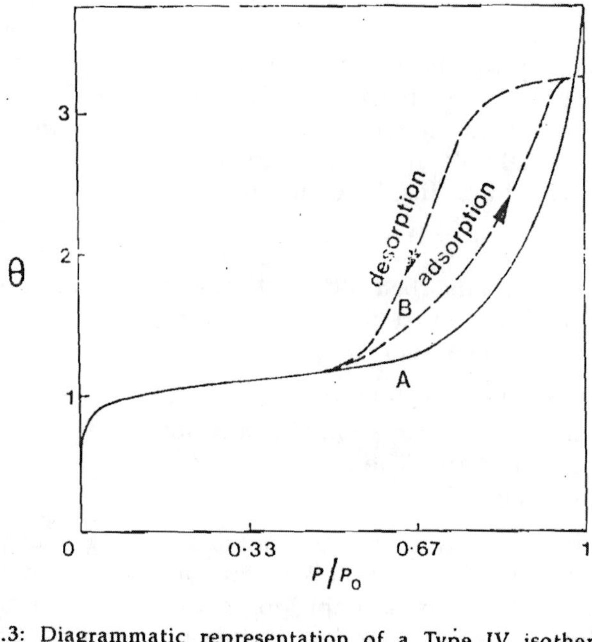

Fig. 30.3: Diagrammatic representation of a Type IV isotherm for adsorption on a mesoporous adsorbent.

The effects of mesopores on the shape of an adsorption isotherm is illustrated diagrammatically in Fig. 30.3, where the transition from a Type II isotherm, A, to a Type IV isotherm B is shown. The two separate braches of the B isotherm correspond to the adsorption and desorption processes, that is to say, the isotherm shows hysteresis. Because of capillary condensation in the mesopores, the uptake at a particular relative pressure will be increased by the amount of condensate in those pores. If pores of a suitable shape are present, then it is possible for capillary condensation to occur below the pressure where the hysteresis loop starts. The origin of the hysteresis loop has been the subject of considerable debate. There are essentially two primary theories. The first is that the radius of curvature of the liquid surface in the capillary differs as the capillary fills from that as it empties as a result of differences normally expected between advancing and receding contact angles. This concept, due to Zsigmondy, has always been open to the objection that these angles are not susceptible to direct measurement.

The alternative ideas assume that hysteresis is due to 'ink bottle' shape pores, such that the wide cross-section of the pores has to be filled before the narrow one, and that during desorption the reverse occurs. If the contact angle is assumed to be zero, then the radius of the pore may be calculated by means of Kelvin's equation, where the radius in this equation, r_k, the Kelvin radius, is related to the pore radius, r_p, by

$$r_p = r_k + t \qquad (8)$$

where t is the adsorbed film thickness. In order to evaluate r_p, then values of t must be obtained from a standard isotherm, either experimental or theoretical.

Pierce has carefully described a calculation scheme for carrying out the derivation of a pore size distribution from a Type IV isotherm. The results of one such calculation are shown in Fig. 30.4.

An alternative way of analysing nitrogen adsorption data at 77.5K on sillica and alunina adsorbents in terms of their pore structure was proposed by de Boer et al, but it requires the use of a standard isotherm determined on a non-porous sample of the adsorbent. This isotherm is represented

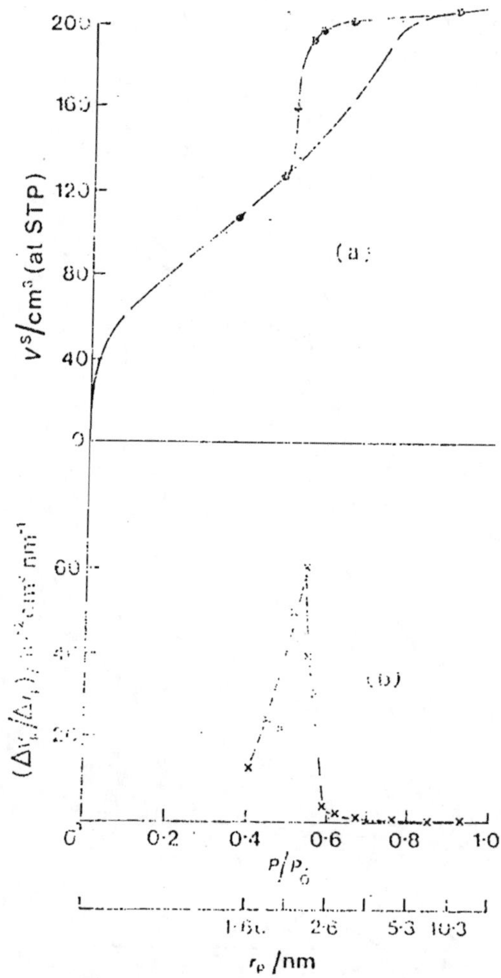

Fig. 30.4: Pore size distribution from the nitrogen adsorption isotherm at 77.5K on titanium (IV) oxide. Data of Harris and Whitaker. *(a)* adsorption isotherm; o, adsorption; o, desorption. *(b)* Pore size distribution curve, V_p and r_p are the pore volume and radius respectively.

not as $\theta = f(P/P_o)$, but as $t = f(P/P_o)$, where t is the statistical thickness of the adosrbed layer calculated using the identity that when $\theta = 1$, $t = 0.354$ nm. Therefore if the experimental isotherm $V^s = f(P/P_o)$ is determined, then a value of t can be found from the standard isotherm, using a particular P/P_0 value. The pairs of values of V^s and t may then be plotted as illustrated in Fig. 30.5.

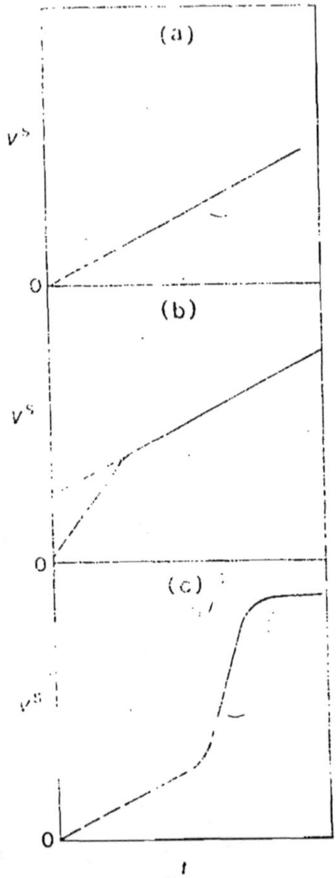

Fig. 30.5: Diagrammatic representation of de Boer 't-plots' for (a) non-porous surface, (b) with some micropores, and (c) with some mesopores.

If we consider Fig. 30.5 (a) for a non-porous surface, then the slope will be proportional to the surface area. If t is expressed in Å and V^s in cm^3 at STP g^{-1}, then the specific surface area may be calculated from

$$s = 15.47 \times \text{slope } m^2g^{-1} \tag{9}$$

Since the standard isotherm of de Boer covers only the region where $P/P_0 > 0.08$, the data are confined to the multilayer region only (θ =1.0at P/P_0 = 0.08), the effects of micropores are usually only experimentally observable at the lowest pressure points of experimental isotherms, and only a few are in the standard isotherm range. Fig. 30.5 (b) shows a t-plot for a microporous surface with the same external surface as in (a), as indicated by the second slope, but with micropores which fill first. The initial slope does not really represent a total surface area, since surface area is a concept which has little real significance for a micropore. However, extrapolation of the higher pressure slope back to the V^s axis gives an intercept proportional to the amount in the micropores, which can be related to the micropore volume by

$$W_o = V^s_{intercept}/22414\rho \ . \tag{10}$$

Fig. 30.5(c) shows the *t*-plot for a solid with the same total surface as in (a), as indicated by the initial slope, but the effects of capillary condensation cause a sudden upturn in the graph, and the final very low slope corresponds to the external surface of the adsorbent which is still available after the capillaries are full. Combinations of types (b) and (c) are not unknown.

Sing has introduced a variation of the '*t*-plot' known as the 'α_s-plot' which essentially avoids using a monolayer capacity in the definition of the standard isotherm, which is instead expressed as V^s/V^s_s, where P/P_0 has some arbitrarily selected value. In practice Sing usually employs $P/P_0 = 0.4$, where it is argued any micropores would have been filled, and capillary condensation has not started. This is useful if the standard isotherm does not possess a sharp knee, and in consequence methods of evaluating V^s_m are likely to be somewhat inaccurate.

Perhaps the most convincing proof of the validity of pore volume analysis lies in what is known as Gurvitsch's rule, which is that although the amount adsorbed at saturation, n^s or V^s, varies from adsorptive to adsorptive, it represents a constant volume when expressed as a volume of liquid, v_1. McKee's results for sillica gel at 25°C are listed in Table 30.1.

The divergence in v_1 is only slightly greater than experimental error.

TABLE 30.1
McKee's results for the uptake at saturation, n^s_{sat}, expressed as a volume of liquid, v_1, for various adsorbates on silica gel at 25°C.

Adsorbate	v_1/cm^3g^{-1}	$n^s_{sat}/m\ mol\ g^{-1}$
Hexane	0.431	3.28
2.3-Dimethylbutane	0.429	3.28
2-Methylpentane	0.431	3.28
Heptane	0.431	2.91
2,2,3-Trimethylbutane	0.420	2.88
Octane	0.434	2.66
2,2,4-Trimethylpentane	0.439	2.63
2,3,4-Trimethylpentane	0.425	2.66
Cyclohexane	0.421	3.88
Methyl cyclohexane	0.425	3.32
Ethyl cyclohexane	0.426	2.99
Benzene	0.440	4.92
Nitromethane	0.449	8.33
Nitroethane	0.434	6.03
Carbon tetrachloride	0.421	4.30